ESTUDIOS e INFORMES de la CEPAL

DEVELOPMENT OF THE MINING RESOURCES OF LATIN AMERICA

UNITED NATIONS

ECONOMIC COMMISSION FOR LATIN AMERICA AND THE CARIBBEAN

Santiago, Chile 1989

LC/G.1355-P
March 1989

This version in spanish has been published with the collaboration of the Department of
Mines of Spain. In its preparation, ECLAC experts were assisted by two experts of the
ADARO National Mining Research Company of Spain S.A. (ENADIMSA): Mr. Carlo
Cuesta and Mr. J.I. García Bernaldo.

UNITED NATIONS PUBLICATION

SALES NO.: E.85.II.G.5

ISSN 0256-9795
ISBN 92-1-121146-8

CONTENTS

CONTENTS OF TABLES

B. Tables in the Statistical Appendix

INTRODUCTION

The evolution of the mining activity in Latin America has been adversely affected by the world crisis, particularly during the period 1980-1983. So, whereas on the one hand, exports of traditional minerals declined as a result of the contraction of world industrial demand and changes in regional consumption levels, which in turn are causing a decline in imports on account of the region's heavy accumulated external debt, on the other hand, it must be borne in mind that the imports of goods produced with mining inputs represent some 40% of the region's total imports and that these resources continue to represent more than 10% of the extra-regional sources of foreign exchange.

In this situation, rapid progress in complementing and integrating the various mining, metallurgical and metal-mechanical phases is needed at the regional level. This requires not only the active participation of the region but also the support of other countries and international institutions for the transfer and adaptation of technologies and adequate complementary financing.

In this regard, a series of joint measures is essential for a better understanding of the mining potential of the region and for a better organization of its production, industrial processing and marketing.

In addition to these measures, Latin America would also be given the opportunity to increase its share of the supply of various mining products to the international market by the year 2000, and increasing its exportable surpluses by becoming more competitive. For this, not only would more joint investment be required but also the opportunity to reach long-term sales contracts which could include industrial complementarity and trade clauses.

In order to identify some of the basic aspects of the scenarios indicated above, this study analyses the trends prevailing in 1984, the effects of the world crisis on those trends, the potential for the future development of the mining resources of the region and puts forward general suggestions which may help in the formulation of the best policies for developing those resources.

SUMMARY

A significant feature of the mining sector is that world production of minerals is highly concentrated with respect to both countries and products. The production of 20 countries including Brazil, Chile, Mexico and Peru represents about 75% of world production while 20 products represent some 95% of the value of world production including eight metallic minerals, and these account for the majority of mining exports from Latin America (iron ore, copper, lead, zinc, silver, bauxite, nickel and tungsten).

During the period 1960-1980 an overall increase in the annual production rates of different minerals was recorded at the world level and this was due to the favourable economic conditions that existed up to the early 1970s. Thus, the annual rates ranged between 2.1% for lead and up to 8.7% for potassium. In Latin America, the trend was similar to the world trend, although it must be pointed out that contrary to what occurred at the world level, potassium had a negative annual rate of -5.3%. The rest of the minerals in the region had annual production rates which, in the majority of cases, were higher than those recorded at the world level.

This view of the evolution of minerals production might appear optimistic if events of the periods 1974-1980 and 1980-1983 are not analysed. During these periods, the world crisis hurt the mining industry and in the majority of cases production rates were negative, as shown in table 7.

The share of the gross domestic product (GDP) of the extractive mining activities in Latin America's total GDP fell from 4.2% to 2.8% during the period 1960-1974. On the other hand, it appears that the mining crisis had less impact on the mining sector than on the rest of the region's economy since its share increased from 2.8% to 3.0% during the period 1974-1982. However, it must be borne in mind that during the subperiod 1980-1982 the mining crisis worsened in the majority of countries of the region.

The evolution of world demand for mining products was uneven at both the regional level and at the product level during the period 1965-1983. A common feature, however, was the presence of downward trends during the subperiods 1974-1980 and 1980-1983. From an analysis of the evolution of the main metallic minerals, it can be established that world consumption grew at annual rates ranging from 1.7% for tin, to 5.6% for nickel, during the period 1965-1974. During the following period 1974-1980, the rates ranged from (-1.6%) for tin and 2.1% for copper. During the last period 1980-1983, the consumption of these minerals was negative except for copper and annual rates fluctuated between (-8.5%) for bauxite and 1.9% for copper. During the period 1965-1980, Latin America's consumption of these minerals showed annual growth rates that were higher than world averages, with figures ranging from 4.6% for tin and 20.9% for nickel during the subperiod 1965-1974 and (-2.8%) for iron and 7.9% for copper, during the period 1974-1980. Between 1980 and 1983, Latin America's annual consumption rates were higher than the world averages for iron (19.9%), nickel (0.0%) and tin (-3.1%) but they were lower for bauxite (-10.9%), copper (13.3%), lead (-8.1%) and zinc (-4.9%).

The changes in the production and consumption structures of minerals, in turn, induced changes in Latin America's share of foreign trade in these products. Between 1970 and 1980, the share of minerals exports fell from 15.2% to 13.54%, increasing to 17.01% in 1983. On the other hand, metals exports showed a downward trend throughout the entire period, with a share of 4.74% in 1970, 3.72% in 1980 and 3.59% in 1983. At current prices, minerals exports grew at an annual rate of 7.9% between 1980 and 1983 after an annual decline of (-1.1%) between 1970 and 1980. This figure was caused by the annual decline of (-7.0%) and (-6.9%) in exports to the United States and Canada respectively. The greatest increases during the period 1980-1983 were achieved in exports to Canada at annual rates of 32.6%, to other developing countries at 12.8% and in intra-regional exports at 7.0%. Between 1970 and 1983, metals exports at current prices suffered an annual decline of (-2.1%) caused largely by the drop in exports to Canada (-30.7%), the countries of the European Economic Community (-6.6%), other developed countries (-3.9%) and Japan (-3.7%). On the other hand, metals exports to the centrally-planned economy countries increased at annual rates of about 21.0% and to other developing countries at 17%.

The impact of the crisis on production structures, consumption and foreign trade in minerals in turn had different effects on relative price levels and for this purpose, the minerals could be classified into three different groups: those which would show up an upward trend in the medium term (1985-1990), those which would maintain an almost constant level and those which would show a downward trend. The first group would include the following minerals: columbium, barite, silver, lithium, magnesium, bauxite, tantalum, tellurium, zinc, vanadium, selenium, chromium and bismuth. the second group would be composed of the following minerals: metallic arsenic, cobalt, gold, ilmenite, fluorite, rutile, nickel, manganese, mercury, copper, phosphated rocks, antimony and cadmium. The minerals with a possible downward trend in their relative price levels would be: thorium, potassium, platinum, molybdenum, sulphur, tin, iron ore, lead, tungsten and asbestos.

The periods 1960-1980 and 1980-1981 were taken as a base in projecting the possible scenario of mining activity by the year 2000. In the period 1980-1981, the main mining situations in Latin America compared with the rest of the world were as follows:

a) the mineral reserves of the region represented more than 30% of the world reserves of niobium, columbium, lithium, iron ore, molybdenum and copper and between 20% and 30% of world reserves of bauxite, selenium, bismuth, nickel and silver. At the other extreme, regional reserves accounted for 1% or less of world reserves of cobalt, gold, potassium, vanadium, chromium, magnesium, platinum and rutile;

b) in the last 15 years, there has been a marked tendency to step up production without at the same time maintaining the original ratio between reserves and production, because of the high costs of prospecting and exploration (risk investments). In Latin America, on the average, this tendency to over-exploit has been more marked than in the rest of the world, in the case of the following minerals: bauxite, bismuth, silver, fluorite, antimony, lead, barite, zinc, tellurium, manganese, tungsten, phosphated rocks, cobalt, gold, vanadium and chromium;

c) although a ratio of 10 to 1 between reserves and annual production is regarded as a minimum acceptable level, this ratio would represent a critical level (less than 10 years) in world averages of asbestos, zinc, tin, lead, tungsten, uranium, magnesium, fluorite, rutile, copper and molybdenum. In Latin America, the reserves with critical levels include the following minerals: gold, barite, zinc, silver, bismuth, chromium, tungsten, cobalt, antimony, lead, tellurium, asbestos, tin, manganese, cadmium and platinum;

d) the minerals in Latin America whose production represented more than 20% of world output included the following: niobium, lithium, bismuth, silver, antimony, bauxite and copper. The minerals with a share of 15% to 20% were: fluorite, tellurium, iron ore, selenium, barite and tin;

e) mineral production in Latin America has a very small share of reserves and output and this shows the relatively minor importance of the region in the industrial processing of its metallic minerals. The production of metals of the region represents between 5% and 17% of world production and includes the following minerals: bismuth, tin, copper, magnesium, lead, zinc, bauxite and nickel;

f) the relatively lower development and the low levels of industrialization in non-metallic mineral processing mean that the consumption of metals and non-metallic minerals from Latin America shows lower ratios than the world average. This consumption represents between 15% and 25% of world consumption of platinum, fluorite and bismuth. The non-metallic minerals representing between 5% and 15% of world consumption were manganese, antimony, asbestos, lead, copper, potassium, magnesium, zinc and mercury;

g) Latin America's extraregional exports were composed of the following minerals (the export percentages compared with their production levels are given in brackets): antimony (72%), bauxite (86%), bismuth (51%), cadmium (40%), cobalt (71%), copper (69%), tin (69%), iron ore (83%), lithium (98%), molybdenum (79%), nickel (76%), niobium (100%), gold (67%), silver (91%), lead (24%), rutile (100%), selenium (91%), tellurium (99%), tungsten (66%), zinc (62%). It must be borne in mind that in terms of value, weight of these products (the ones underlined) represented around 95% of total mineral exports which, in turn, constituted about 10% of the total exports of goods from the region. Nevertheless, this figure was higher in the following countries: Bolivia (59%), Chile (53%), Guyana (44%), Jamaica (76%), Peru (34%), Dominican Republic (13%) and Suriname (82%);

h) in 1980, Latin America's extraregional imports were composed of 12 products, and the following seven accounted for 99% of the cost of the year's mineral imports with the following proportions of total supply: asbestos (57%), barite (6%), chromium (16%), fluorite (1%), magnesium (1%), manganese (5%), mercury (79%), platinum (99%), potassium (86%), phosphated rocks (52%), uranium (47%) and vanadium (22%).

The projection of the possible scenarios of mining activity in the year 2000 were based on the following assumptions:

a) that depending on the individual case, the per capita consumption of the developed countries would increase to annual rates of 0.5% to 1%, given their high saturation levels. The per capita consumption levels of the centrally-planned economy countries would increase at rates which, depending on the product, would mean that they would reach 75% to 100% of the levels reached by the developed countries during the base year, 1980. For their part, the levels in Latin America would reach 25% to 100% of the levels reached by the developed countries in 1980. The per capita consumption of the other developing countries would grow more rapidly than those of the other regions so that it would at least be higher than Latin America's 1980 levels. On the basis of these assumptions, world mineral consumption would grow at annual rates which would fluctuate between 5.0% (chromium) to 7.8% (vanadium) in one group of 16 products, where in the second group of minerals it would range from 2.0% (tin) to 4.7% (antimony);

b) that in the first place, the production of each mineral at the world level would be equal to its consumption. In the second place, it was assumed that each group of countries would achieve a high degree of self-sufficiency, the only limitation being that the estimated reserves for 1981 could not be exhausted before the period 1995-2000.

14

Consequently, the production of the developed countries would have annual growth rates ranging from (-0.2%) for magnesium to 10.2% for cobalt. The centrally-planned economy countries would have an annual production growth rate of between (-0.01%) for asbestos to 19.2% for rutile. The production of the other developing countries would fluctuate between (-4.3%) for silver and 29.3% for magnesium. In Latin America, the various ranges would include the following minerals:

    i) lithium, uranium, vanadium, cobalt, rutile, phosphated rocks, mercury, molybdenum and potassium with annual growth rates above 10%;

   ii) platinum, iron ore, selenium, nickel, asbestos and cadmium with rates between 5% and 10%;

  iii) fluorite, bauxite, manganese, tungsten, lead, antimony, tellurium, tin, chromium, bismuth, copper, silver and zinc with positive rates below 5%;

  iv) barite and gold with negative rates (to avoid exhausting reserves before the period 1995 to 2000).

In order to achieve the above rates, the annual rate of total mining production would have to increase to 3.6% at constant 1982 values;

c) that the differences between production and consumption in each group of countries would determine the exportable surpluses or the regional import requirements. In Latin America, exportable surpluses would be created for the following: antimony, bismuth, cadmium, copper, tin, fluorite, iron ore, lithium, molybdenum, silver, selenium and zinc which would be shipped primarily to the other developing countries and the centrally-planned economy countries. In turn, the region would have to import asbestos, barite, cobalt, chromium, ilmenite, manganese, mercury, gold, platinum, rutile and tungsten which would be supplied primarily by the group of developed countries and the centrally-planned economy countries (see table 21);

d) that the production-cost-price ratios established during the period 1947-1974 would remain constant and the forecast for minerals price indexes in the year 2000 was made on that basis. According to this assumption, the prices of cobalt, tellurium, platinum, rutile and uranium would increase more rapidly while the prices of asbestos, cadmium, bismuth, antimony and nickel would increase more slowly;

e) that as a variant of the above scenario and considering the large mining resources potential in the region, the possibility of import substitution is being contemplated with a view to achieving self-sufficiency in minerals at the regional level;

f) that considering the production and import substitution levels, a total investment of over US$ 64 billion at 1975 prices has been estimated. This amount could be concentrated during the first ten years of the period and therefore, US$ 6.4 billion would be required annually exclusively for the production of minerals and concentrates.

It is expected that, at the end of the period, external financing would represent 20% of that investment;

g) that by the year 2000, the share of extractive mining in the total output of the region would be estimated at 2.6%, whereas it is expected that the output from metallurgical and metal-mechanic activities would increase more rapidly. Mining exports would continue to represent around 10% of the exports of goods from the region and their net contribution to foreign exchange earnings has been estimated at more than US$ 3 billion per year (see table 24).

Chapter I

THE INTERNATIONAL MINERALS MARKET: FEATURES AND OUTLOOK

1. Formation of the North Western macro-market and the international division of labour

After the Second World War, the United States economy, where the bulk of international capital was concentrated, had no similar counterpart for absorbing its surplus output, its production capacity having increased during the war. In order, therefore, to widen its own scope for expansion, the United States began to assist in the reconstruction of the European and Japanese economies, through vast economic assistance and direct investment programmes and it did so, mainly by opening up its market to the new European and Japanese production. That was how the so-called "macro-market of the northern hemisphere" began and it shaped a new international division of labour through industrial specialization and a large volume of trade among the economies. Thus, the annual growth rate of trade between these countries, which between 1928 and 1938 stood at 1.5%, rose to 11.7% during the period 1950-1975. It may be that for lack of long-term planning, the industrial structures of these countries ceased to complement one another, reached a high level of self-sufficiency and gradually began to compete with one another. The result was to reduce any further chances of developing the macro-market of the North. In fact, its trade grew by only 3.8% from 1973-1979 and by 2.8% from 1973-1980. This situation may be bringing about the demise of a development model of the northern macro-market and consequently the pattern of international division of labour implicit in that model. This situation could be one of the major causes of the present world crisis and success in overcoming the crisis may well hinge on the implementation of new development models and the establishment of a new pattern for the international division of labour or a series of patterns with a high degree of self-sufficiency at the regional level or among large groups of countries.

At the end of the 1960s and in the early 1970s, the high level of mass consumption in the developed countries exerted a strong pressure to obtain higher wages, with the result that profitability gradually declined. This combination of facts may have caused a decline in the investment rate of enterprises which retaliated by pushing prices up in an attempt to recover their profit margins. This may have given rise to the phenomenon of stagflation. From 1973, the inflationary process that had been taking place deteriorated even further when the Organization of Petroleum Exporting Countries (OPEC) imposed its oil pricing policy. The inflation rate, which during the period 1966-1973 was 3.7%, soared to 7.8% in 1973 and to 13.4% in 1974. Recessive measures taken by various governments to control this inflation led to a further reduction of any chances of investment in the developing countries and triggered a rapid process of internationalization in which the resources of international financial system were channelled to the developing countries, where wage levels were still lower than in the industrialized countries.

To ensure that profit margins were maintained, the transnationalization process necessitated harsher external debt conditions (which had a positive effect on the terms of trade) and the concentration of the economic surplus primarily in the national financial systems. Keeping the interest rates on the financial resources high is not only imposing additional curbs on investments in the developed countries but is also depriving the developing countries of new resources on account of the already high cost of servicing the external debt.

Like the rest of the developing world, Latin America has been kept out of the mainstream of the industrialization process of the macro-market model of the North and its role in the international division of labour has been confined to exporting agricultural and mining raw materials and importing manufactured goods, a situation which has hampered the region's development. In truth and in fact, the essence of Latin America's economic thinking derived from its repudiation of the situation of inequality produced by the new international division of labour. Relations between the developed countries, at the centre and the developing countries at the periphery, tended to accentuate the economic disparities between both groups even more, because of the constant deterioration of the terms of trade. Industrialization based on import substitution at the national level was held up as an alternative model.

In order to break the restriction of the limited national markets, the so-called balanced growth theory was formulated and subsequently strengthened by the theory of integration among the countries of the region through the establishment of the corresponding operating mechanisms (LAFTA, CACM, Andean Pact). According to the situation of each country, the industrialization process evolved with very specific features and achieved some success in the region as a whole in that total output grew at an annual rate of 5%. However, the failure to achieve either balanced growth or to accelerate the pace of integration, coupled with a repetition at the national level of the conditions of concentration of income and of centre-periphery relations, created an exceedingly inequitable productive structure and rapidly exhausted all effective substitution options.

Furthermore, the concentration of resources in the substitution process led to a decline in the export rate and reduced the capacity to purchase imports, whose ratio to output fell from 17% to 10%. The growing demands of this industrialization process for imports of machinery, equipment, spare parts and intermediate products and the foreign exchange restrictions subsequently triggered off the crisis of the present model causing deep disappointment over the failure of the national autonomous development policies.

During the next period (1965-1974), Latin America tended to apply a new development strategy using an outward growth model, which was made easier by greater import demands from the developed countries. Some of its features were different from those of previous periods: rapid export expansion, more favourable terms of trade, greater access to direct investments and to official sources of credit, a larger share in the exports of manufactured products which came to account for as much as 15% of the value of total exports and an increase in the ratio of the external purchasing power which rose from 10% to 16%. Under those circumstances, the region achieved substantial economic growth, with unprecedented annual growth rates above 6%. The situation in Latin America changed dramatically during the period 1980-1983, when both export growth rates and raw materials prices fell; semi-processed and manufactured goods were now faced with new protectionist policies while import prices increased.

During the period 1970-1980, internationalization of the capital flow to the region increased and net indebtedness grew to an annual rate of 22%. This heavy flow of financial resources enabled the region to sustain a high import demand which in the final analysis affected internal production levels, for despite the magnitude of external resources, the growth rates of output were lower than those of the previous period: 5.6% between 1975 and 1980, 1.5% in 1981 and negative rates in 1982 and 1983. On the contrary, the harsh external credit conditions determined that debt servicing increased at an annual rate of 13% (1975-1982), and this situation sparked off the worst financial crisis in Latin America's history.

2. The main features of the world mineral prices

The 1973-1974 oil crisis was in fact one phase of another deeper crisis, the seeds of which were sown during the Second World War and have continued to grow up to the present. That was the crisis of mining raw materials, the basic features of which are outlined below. In the first place, profound changes were occurring in the world political structure with the inclusion of a number of countries which on gaining political independence took stock of their economic and social needs and all they had to meet those needs were some natural resources including some mining resources that were needed by the developed countries. However, trade between both groups of countries had not developed fairly and this only widened the gap between them. Faced with this situation, the developing countries initiated a series of measures to strengthen their sovereignty over their resources and to obtain a greater share of the profits generated by the exploitation of these resources. These measures ranged from nationalization of foreign enterprises operating in the different countries to the establishment of State enterprises.

After the period of recovering sovereignty over their resources, producer countries began to give consideration to forming groups to defend their common interests, to demand better distribution of mining income and established models for operating in the international minerals market. For their part, the consumer countries took various measures which have enabled them to reduce their dependency on minerals inputs or at least to ensure them a regular supply. Some of these measures included the following: promoting the increase of national production, encouraging the conservation, or limiting the use of minerals, increasing the substitution and recycling processes of the most critical minerals, increasing the exploration and exploitation of marine mining resources, establishing strategic reserves as inventories and securing appropriate diversification of the external sources of supply of these products inter alia, through long-term sales contracts.

The various scenarios, which have been taking shape, have spurred a sudden change in the world market structures among the countries. The inequitable supply-demand ratio which exists between consumer and producer countries, combined with the somewhat latent geopolitical factors, have produced a chain of cause and effect which has upset the difficult world minerals market.

On the other hand, it must be realized that there are some rather homogeneous factors which, overall, dictate the manner in which the market usually operates, for example:

    i) physical determining factors affecting the possession and availability of mining resources;

    ii) environmental determining factors relating to the restrictions currently being imposed on mining activity in order to protect the environment;

    iii) determining factors of technological substitution;

    iv) political-commercial decisions affecting the structure of the world market;

    v) economic determining factors affecting the ad hoc fluctuations in demand and price stability.

a) <u>Physical determining factors</u>

The size and distribution of non-mineral reserves combined with access to them, are obviously the main physical factors which determine potential world mineral production. Now, in the case of non-renewable resources, those factors generally impose a limit and it must be determined how far and within what time-frame that limit is likely to become a restrictive or critical element in the potential supply.

The topic of limits on mining resources has been updated in two reports of the Club of Rome published in March 1972 (D. H. Meadows, "The limits to growth") and in October 1974 (M. Mesarovic, E. Pestel, "Second report of the Club of Rome"). Other reports published subsequently also tried to define the restrictions of renewable net resources in the evolution of economies. Of these, the following should be highlighted: the report of the United Nations headed by W. Leontief and published in 1977, the OECD report published in 1979 entitled "Inter-futures" and "The global 2000 report to the President" commissioned by President Carter and published in 1980. All of these studies attempt, in one way or another, to describe the world at the end of this century in terms of a set of previously analysed parameters.

In the Leontief study, the pessimistic natural resources option concludes that by the year 2000 the group of developed countries would face critical production limits of <u>nickel</u>, <u>zinc</u> and <u>lead</u>. In the group of developing countries, this would occur in the case of <u>copper</u>, <u>zinc</u> and <u>lead</u>, whereas in the centrally-planned economy countries this would occur in the case of <u>copper</u>, <u>nickel</u>, <u>zinc</u> and <u>lead</u>. In the report entitled "Facing the future" (<u>Interfuturos</u>) a forecast is also made of the mining resources that could be regarded as reserves, because of the considerable increases in their prices in the medium and long term. Generally, it is observed that the ratio of resources to reserves would be 3:1, but it is determined that by the year 2000 there would be critical production problems worldwide in respect of <u>bismuth</u>, <u>mercury</u>, <u>lead</u>, <u>zinc</u>, <u>asbestos</u> and <u>silver</u>.

The report commissioned by President Carter concludes that "the relatively short life expectancy of certain substances does not imply that they would be exhausted immediately but it does clearly indicate that the reserves of at least half a dozen minerals --<u>industrial diamonds</u>, <u>silver</u>, <u>mercury</u>, <u>zinc</u>, <u>sulphur</u> and <u>tungsten</u>-- must be increased in order to sustain adequate production levels in the decades to come".

According to the figures on reserves in 1981 and production in 1980 published in 1982 by the Federal Institute for Geosciences and Natural Resources of Hannover (Federal Republic of Germany), it is estimated that at the world level, <u>asbestos</u>, <u>lead</u> and <u>zinc</u> reserves would reach critical limits. In Latin America, this situation would occur in the case of asbestos, chromium, cobalt, manganese, platinum, tungsten, vanadium and zinc.

On the basis of those forecasts, there is reason to believe that there are areas or regions whose high level of supply and access to certain mining resources places them in the best position to make full use of them, with the opportunity to trade with other regions that have other resources, capital goods or technology. This theoretical division of mining production on a worldwide scale would reduce the medium-term needs for financial resources for mining resources prospection and exploration, except those earmarked for the production of those which are scarce, worldwide, such as lead and zinc.

b)   Environmental factors

Some environmental pollution is caused by extraction and processing of minerals with serious effects upon the earth, water and atmosphere and some by the metallurgical industry. It is highly likely that, in the future, the ratio of mineral to sterile material and metal to mineral will continue to deteriorate, exacerbating environmental pollution problems even more. Furthermore, in the region, a significant amount of mining is done in areas with a fragile ecological balance in which replacement work can be very difficult and costly. The possibility of marine exploitation of mineral resources brings with it also possible problems of polluting the continental shelves and international waters. These considerations have informed the legislation on environmental protection in various countries imposing on mining activities a series of restrictions and additional costs which represent between 10 to 20% of the net income of many projects, making their exploitation uncompetitive internationally, and reducing the world mineral supply even more.

c)   Determining factors of technological substitution

In earlier periods, one mineral was usually substituted for another or for another type of product in the hope that in medium term, there would be a great difference in their relative prices or that better qualities would be found in the substitute product. In this context, lower production of a given mineral is compensated to a certain extent by increased production of the substitute mineral. In recent years, there has been a noticeably strong trend towards technological substitution caused by the shift to industrial structures, which make even greater use of high technology and service industries. This process can take various forms, such as:

- Quantitative substitution which presumes the reduction of the metallic input per unit of the end product. For example, the higher cost of hydrocarbons is determining how vehicles are manufactured and consequently there is a reduction of at least one-third of the metallic components.

- Substitution of production procedures, which input a smaller quantity of metals, such as the changes taking place in the electronics industry in the production of mini-components.

- Functional substitution which presumes that large production or services systems will be replaced, as for example, the changes that are taking place in the communications, electronic data processing and mini-processing systems. Obviously, wherever there is substitution, the curbs on the production of component minerals of these industrial activities would not be offset by increases in the production of other minerals and this constitutes a permanent limitation.

d) **Political-commercial decisions**

The united stand of the developing countries has enabled them to defend their interests in international fora. Thus, the declaration made at the Tenth Special Session of the United Nations General Assembly, held in 1974, was embodied in a programme which summarizes their aspirations known by the generic title of the New International Economic Order (NIEO). Subsequently, both the developed countries and the OPEC countries refused to have any dialogue on the subject of the NIEO and insisted that international trade should be regulated by the economic laws of a competitive market and that multilateral decisions should continue to be taken within the framework of the appropriate organizations: the GATT, the World Bank and the International Monetary Fund. In 1976, UNCTAD absorbed a part of this programme, adopted the "Integrated Programme for Commodities" which includes the negotiation of a series of agreements for a selected group of resources. The fundamental element of the Integrated Programme is the so-called Common Fund, which is an instrument designed to meet the financial requirements deriving from the operation of the agreements. Both producer and consumer countries would participate in these agreements in an attempt to structure the market of each resource. However, because no concrete results have been achieved, the aspirations of the developing countries are once again being frustrated. The Cancún Conference held in Mexico in 1981 met with the same fate; however, France put forward a suggestion, which was supported by Canada and Sweden, that the developed countries should discuss a far-reaching allocation of resources and a technical co-operation plan with the representatives of the Group of 77. This suggestion was strongly opposed by the United States and the United Kingdom and for various reasons, Japan, the Federal Republic of Germany and the OPEC countries also adopted the same position.

This crisis situation led the major importing countries to establish a series of measures intended, on the one hand, to minimize their use of any mining resources that they do not possess and on the other, to reduce their dependence on any imported supplies that could be cut off. The exporting countries, in order to deal with the urgent balance-of-payments problems facing most of them, have taken various measures to maximize their income from mining exports such as the establishment of voluntary production and export quotas in order to maintain or increase the price of their products. In the United States the heavy reliance of industry on the imports of mining products such as antimony (51%), asbestos (80%), bauxite (94%), cobalt (91%), chromium (90%), tin (80%), manganese (98%), nickel (72%), silver (50%), potassium (68%), tungsten (52%) and zinc (67%), is causing serious concern. A series of measures has therefore been proposed, including the following (see table 18 in the Statistical Appendix):

a) The promotion of investment abroad in its own or shared programmes and projects, provided that this does not hurt the domestic mining interests of the country. The main incentives are: to reduce or remove double taxation, to reduce import duties and to develop international arbitration procedures in nationalization or expropriation disputes.

b) The maintenance of close trading relations with South Africa, Australia, Canada and Mexico, countries on whose minerals the United States relies heavily.

c) The strengthening of its naval power to protect important seaways, especially those connected with the transportation of hydrocarbons.

d) The implementation of specific actions under the Trioceanic Alliance, which is understood to be a combination of the NATO countries plus South Africa, Saudi Arabia, Australia, Brazil, Egypt, Indonesia, Mexico, Nigeria, Singapore and Zaire. This alliance will not only achieve greater military and economic superiority but also a high concentration of scientific knowledge, food supplies, oil reserves and the most important minerals.

The European Economic Community is another region that is heavily dependent on foreign sources for minerals, as shown in the following structure: 20% of its needs are met from own resources, 40% from the resources of other developed countries and 40% from the resources of the developing countries. With respect to the main products, its import needs would represent the following percentages of its total supply: alumina 84%, antimony 91%, asbestos 82%, cobalt 100%, copper 67%, chromium 100%, tin 95%, iron ore 79%, manganese 99%, mercury 86%, molybdenum 100%, nickel 80%, gold 99%, silver 98%, lead 45%, phosphated rocks 99%, tungsten 77%, vanadium 100% and zinc 52%. It would appear that the Community does not have much scope to increase self-sufficiency in these minerals since it is the oldest consumption centre and has almost exhausted its resources. High dependency could therefore be tolerated if the EEC manages to achieve a high diversification of its supply sources; however, while direct European investment has remained stable for many years, it has not reached the levels of the United States and Japan. Furthermore, these investments have been directed primarily towards other developed countries, which are politically more stable, but which also have higher indexes of auto-consumption. Because of this, the European Community is proposing to undertake a series of joint measures including the following:

a) In its first document to the Council, in 1975, the Commission analysed the risks to Europe of high dependency on supplies of primary raw materials from the third world countries and suggested basic guidelines for developing a Community policy in this area. This line of action is centered on secure long-term supplies, the need to guarantee mining investments abroad, price stabilization, the possibility of increasing the region's mining resources and the use of economies of scale in the industrial processing of those raw materials.

b) In 1978, the agencies of the Community declared the mining sector a priority area and proposed another series of joint measures which were also intended to promote the exploration and production of its own resources and to facilitate the entry of mining resources from abroad. These measures include:

i) The implementation of a multi-annual programme of research and development of its own resources (1978-1981). This programme was later extended to the period 1982-1985, and the basic aims of exploration, processing and mining technology continued to be the focus of attention.

ii) In January 1978, the Commission also submitted to the Council guidelines for Community action in the field of research in the developing countries. These guidelines basically referred to the activities of its mining enterprises abroad in respect of exploration and investments within the framework of promoting and safeguiding those investments.

c) The establishment of the Lomé Convention. So far Lomé I and Lomé II have been signed and Lomé III will be signed with 66 countries of Africa, the Caribbean and the Pacific (ACP) early in December 1984.

i) The STABEX system which was implemented under the first Convention is designed to offset the effects of a sudden drop in the export earnings of the ACP, through a financial transfer from the European Development Fund (EDF) and which will cease when the original position is recovered. This system includes iron ore.

ii) SYSMIN is the system which covers the other minerals and was instituted in Lomé II. It provides the ACP group of countries with the minimum protection that is indispensable to maintain and develop their export capacity in the case of natural disasters, grave political events or a drop in prices.

SYSMIN covers exports of copper and cobalt from Zambia, Zaire and Papua New Guinea, phosphates from Togo and Senegal, manganese from Gabon, bauxite and alumina from Guinea, Jamaica, Suriname and Guyana, tin from Rwanda and pyrites and iron ore from Mauritania and Liberia.

iii) With the aim of developing the mining potential of the ACP group of countries, the European Investment Bank (EIB) provides the appropriate technical and financial assistance through long-term loan agreements.

Japan is another developed country which relies very heavily on external sources for its supply of minerals. Its import requirements represent the following percentages of its total supplies of the major minerals: antimony 100%, asbestos 99%, bauxite 100%, cobalt 100%, copper 87%, chromium 99%, tin 96%, iron ore 99%, manganese 97%, molybdenum 99%, nickel 100%, gold 94%, silver 73%, lead 75%, phosphate rocks 100%, tungsten 75%, vanadium 100% and zinc 59%. In order to obtain or secure a regular supply at the lowest possible cost, Japan has pursued a policy of diversifying its external sources of minerals, in the context of international market conditions; these sources are located primarily in the countries of the Pacific and Asia, Australia, South America and exceptionally in countries of Africa. This policy is applied on the basis of two main instruments: the signing of long-term sales contracts tied to loans for mining equipment and direct participation (joint ventures) in projects or mining enterprises.

The USSR is the second world producer of minerals after the United States and possesses huge reserves of non-energy minerals and plays such an important role in the international market, that its minerals, metal and energy exports represented 50% of the total value of its exports in 1979. In recent years, however, these have fallen significantly while at the same time, imports from the group of centrally-planned economy countries have increased, especially imports of chromium, tin and lead. It is estimated that the USSR's import requirements for the main minerals would have the following percentages of the country's total requirements: antimony 20%, barite 50%, bauxite 50%, cobalt 43%, tin 11% and fluorite 47%.

Attempts by the minerals producing countries to control their export markets have met with little success so far. These attempts seek to attain the following main objectives: maintaining or increasing prices in situations of surplus supply which are produced during periods of low economic expansion in the industrialized countries; avoiding further deterioration of the terms of trade with the developed countries; maximizing foreign exchange earnings to face their balance-of-payments problems. The formation of producers' associations (CIPEC, IBA) on the model of OPEC was a more serious attempt to unite or combine the interests of the producer countries. However, for various reasons, including changes in the aims with which they were created and the lack of political agreement on their concepts and procedures, these organizations have not been as effective as hoped.

In addition to the possibility of forming oligopolies among producer countries, it must be borne in mind that in the minerals market, on the one had, there are oligopolic forms at the level of transnational enterprises with an increasing tendency to integrate the production of a growing number of minerals horizontally, including those which can be substituted among themselves and on the other hand, those in which vertical integration still persists in the form of successive phases of the production process and industrial processing of the different minerals. These enterprises are obviously acting in pursuit of their own interests, which in some cases and circumstances may coincide with the interests of the producer countries, in which case they could pursue joint action, for example, to support prices and determine the distribution of marginal income. Apart from this type of enterprise, there is another type of international company which markets only mining raw materials and exercises a certain degree of influence and control over these markets. The world crisis has served to strengthen the role of the enterprises and improve their profitability and are becoming one of the most dynamic agents in the minerals market. Producer countries could therefore be associated with the existing ones or form organizations of this type.

While the trade policies of the importing and exporting countries could create a measure of stability in the minerals market, they would not be enough to overcome the problems of the minerals crisis, which would only serve to worsen the economic frustration of the developing countries, with the subsequent social conflicts and political instability.

e)   Economic determining factors

In view of the above circumstances, there is no doubt that the price of minerals plays a most important role in regulating international trade, and in fact the lack of natural resources is more a problem of prices than of the physical availability of supplies. Nevertheless, it must be remembered that price fixing also presents technological and geopolitical limitations which restrict or hinder the success of such a mechanism.

The wide and frequent price fluctuations derive from imbalances created in the supply-demand ratio, since the elasticity of demand in those same products is relatively unresponsive to the more rapid variations in demand. The slow response of the mining sector to a sudden industrial reactivation or vice versa makes prices register significant and in some cases extreme variations and this creates marked distortions.

Table 1

EFFECTS OF THE CRISIS ON THE VOLUME OF WORLD PRODUCTION OF MINERALS

(In percentages)

	Period 1947-1974		Period 1974-1982	
	With high growth rates	With low growth rates	With positive growth rates	With negative growth rates
I. Traditional non-ferrous metals				
	Columbium 8.5 a/	Copper 4.8 a/	Thorium 6.6	Copper -0.2
		Zinc 4.7	Columbium 1.0	Zinc -0.5
		Lead 3.8	Tin 0.4	Lead -1.4
		Tin 2.7		Metallic arsenic -7.9
		Metallic arsenic -0.6		
II. Traditional ferrous metals				
	Vanadium 11.1	Tungsten 3.8	Vanadium 3.8	Manganese -0.2
	Molybdenum 7.3	Tantalum 0.4 a/	Chromium 0.4 a/	Iron ore -1.8
	Iron ore 7.0		Tungsten 2.4	Tantalum -2.6
	Nickel 6.9		Molybdenum 0.7	Nickel -3.2
	Manganese 6.5			Cobalt -3.2
	Cobalt 5.8			
	Chromium 5.3			
III. Refining - insulating ores				
	Fluorite 7.5			Fluorite -0.8
	Asbestos 6.5			Asbestos -0.8

Table 1 (contd.)

	Period 1947-1974		Period 1974-1982	
	With high growth rates	With low growth rates	With positive growth rates	With negative growth rates
IV. Electric use metals				
Selenium		5.0 a/		-0.1
Cadmium		4.7		-1.6
Tellurium		3.2 a/		
Mercury		2.0		-3.3
Tellurium				-8.9
V. Chemical use minerals and metals				
Potassium	9.0			-0.1
Phosphated rocks	7.3		0.4	
Sulphur	6.7		10.8	
Bismuth		4.4		-4.8
Barite		4.1	6.0	
Antimony		2.4		-3.6
Lithium				-15.9
VI. Light metals				
Bauxite	9.8			-0.4
Ilmenite	9.5		1.0	
Rutile	9.3		0.5	
Magnesium	7.7		8.2	
VII. Precious metals				
Gold	9.7		0.8	
Silver	9.7		3.0	
Platinum			1.4	

Source: See table 1 of the Statistical Appendix.

a/ Period 1965-1974.

Three situations are observed in the evolution of metal prices: a long-term trend, cyclical fluctuations of the market and short-term fluctuations. The long-term trend is so described because its changes are slow and regular and indicate a movement towards a balance between mining production and consumption. The cyclical fluctuations respond to the phenomena of adjustment or imbalance between supply and demand, which react to situations in the economies of the consumer countries and are affected by the start-up of previously inactive mines. The short-term fluctuations are due more often to speculative phenomena which are generally caused by forces exogenous to production and consumption in the strict sense of the term.

3. Evolution of production and world consumption of minerals

As can be observed, the evolution of world production, during the period under review, has not only been closely linked to the dynamism of the domestic industrial sectors which have for the most part affected the levels and trends of minerals consumption, but also to the evolution of the international economy. This evolution affects the levels and trends of foreign trade and therefore, the movement of international prices.

In the period 1947-1974, all the minerals except for metallic arsenic had positive annual growth rates of production. As a result of the world crisis, during the period 1974 to 1982, all the products with the exception of sulphur, barite, magnesium and silver, saw a decline in the growth rates of their production, in some cases even reaching positive values while in others, the majority of them showed negative rates (see table 1).

In the subperiods of the world crisis: 1974-1978 and 1978-1982, the following developments in physical production were observed:

i) An increase in production preceded an evolution in which both demand and prices were favourable.

- During the subperiod 1974-1978 the minerals in question were: asbestos, barite, bauxite, ilmenite, tin, magnesium, manganese, molybdenum, silver, platinum, thorium, tungsten and vanadium.
- During the subperiod 1978-1982 the minerals concerned were: barite, copper, columbium, tin, lithium, mercury, gold, silver, platinum, potassium, rutile, vanadium and zinc.

ii) A decline in production (quotas), in order to maintain or increase price levels in the face of undynamic demand.

- During the subperiod 1974-1978 the minerals in question were: arsenic, cobalt, columbium, fluorite,

iron ore, lithium, nickel, gold, lead, potassium, rutile, tantalum and tellurium.
- In the subperiod 1978-1982 the minerals concerned were: arsenic, asbestos, sulphur, bauxite, cobalt, fluorite, iron ore, ilmenite, magnesium, manganese, nickel, phosphated rocks and thorium.

iii) A decline in production, in view of a negative evolution of demand and prices.
- In the subperiod 1974-1978 the minerals in question were: antimony, bismuth, cadmium, copper, mercury and zinc.
- In the subperiod 1978-1982 the minerals concerned were: antimony, bismuth, cadmium, chromium, molybdenum, selenium, tantalum, tellurium and tungsten.

iv) Increases in production despite the decline in prices because of the negative or undynamic demand.
- In the subperiod 1974-1978 the minerals in question were: sulphur, chromium, phosphated rocks and selenium.
- In the subperiod 1978-1982 the mineral involved was lead.

Taking 1974 as the base year, the physical production index of all the minerals analysed increased to 104 in 1978 with an annual growth rate of less than 1% between both years. The index for the year 1982 shows that the physical production of that year was lower than that of the base year 1974 (index = 98) with a negative rate of -3.2% in each of the years during the period 1978-1982 and -0.25% during the period 1974-1982. At the same time, the price index for this group of minerals went from 137 to 176 for the years 1978 and 1982, in other words, prices increased at the annual rates of 8.19% during the period 1974-1978 and 6.46% during the period 1978-1982. As a result of the evolution of the physical production and of prices, the index of the value of production (incomes) was 143 for the year 1978 and 173 for the year 1982, with annual growth rates of 9.35% during the period 1974-1978 and 4.88% during the period 1978-1982.

If the annual growth rates of income or value of production of the first period of the crisis 1973-1978 are compared with those for the evolution of the previous period, 1950-1973, it will be observed that there was a positive evolution in 50% of the products analysed and a lower or negative evolution in the other 50% (see table 2). In view of the fact that the main export minerals in the first group are bauxite, silver and tungsten and in the second group, copper, iron ore, nickel, lead and zinc, this would indicate the need for Latin America to have a more diversified export structure, more responsive to the variations in demand and prices. Compared with the income levels between 1974 and 1978, it is observed that the products which had greater relative expansion were: cobalt, arsenic, molybdenum, columbium,

Table 2

EFFECTS OF MINERALS ON THE VALUE OF PRODUCTION
(In percentages)

	Period 1950-1973		Period 1973-1978	
	High growth rates	Low growth rates	High growth rates	Low growth rates
I. Traditional non-ferrous metals				
	Copper 7.9	Zinc 3.7	Columbium 5.6	Lead 3.1
		Lead 2.2		Zinc -3.3
				Copper -10.2
II. Traditional ferrous metals				
	Vanadium 18.0	Iron ore 5.3	Cobalt 24.8	Tantalum 5.0
	Nickel 9.6	Cobalt 4.9	Chromium 23.1	Vanadium 2.5
	Molybdenum 7.9	Tungsten 4.1	Tungsten 22.7	Iron ore 2.2
		Chromium 2.1	Molybdenum 18.0	Nickel -4.2
		Manganese 0.8	Manganese 7.6	
III. Refining, insulation and other minerals				
	Feldspar 7.6	Kaolin 5.8	Asbestos 17.5	Feldspar 3.6
		Asbestos 5.3	Kaolin 7.9	Fluorite -0.2
		Fluorite 5.1		

Table 2 (concl.)

	Period 1950-1973		Period 1973-1978	
	High growth rates	Low growth rates	High growth rates	Low growth rates
IV. Metals of electrical use		Mercury	5.4	Mercury −23.7
V. Minerals and metals of chemical use	Sodium 8.2	Phosphated rocks 5.9	Sodium 20.0	Antimony −1.4
	Borate 7.2	Barite 4.6	Phosphates 19.8	Nitrates −5.9
	Potassium 6.2	Sulphur 4.0	Sulphur 13.2	
		Antimony 1.6	Barite 12.8	
		Nitrates −4.5	Potassium 8.7	
VI. Light metals	Rutile 11.9	Magnesium 5.2	Magnesium 22.6	Ilmenite 0.6
	Bauxite 8.1		Bauxite 14.3	Rutile −2.9
	Ilmenite 6.0			
VII. Precious metals and stones	Platinum 10.5	Silver 5.4	Silver 10.1	Industrial diamonds 4.8
	Industrial diamonds 8.7	Gold 3.9	Platinum 6.6	Gold 4.2

Source: See table 2 of the Statistical Appendix.

tellurium, thorium, tungsten, chromium, sulphur, phosphated rocks, tin, barite, bauxite and magnesium. On the other hand, those with a negative evolution were: selenium, copper, antimony, cadmium, zinc, mercury and nickel. During the period 1978-1982, the products with the greatest relative expansion in their incomes or their total values were: columbium, sulphur, barite, thorium, magnesium, arsenic, platinum and gold whereas those with a negative evolution were: lead, copper, tellurium, antimony, cadmium, selenium and bismuth.

As seen in table 8 of the Statistical Appendix, there was a positive evolution in the consumption of the major minerals during the period 1965 to 1980, with the exception of tin which after a growth rate of 1.6% during the subperiod 1965-1974, began to show negative growth in the following two periods. However, from the year 1980, the growth of minerals consumption stagnated due to the cumulative effect of the world economic crisis which had a serious effect on mining, even though this occurred later than in the other branches of industrial activity.

During this last period, copper consumption maintained a positive growth rate of 1.9%. The greatest rate of decline (-8.5%) was recorded in bauxite during 1980-1983. The other minerals maintained a negative growth rate ranging from (-6.4%) to (-1.1%).

At the regional level, there was a gradual decline in the developed countries' share of consumption and a corresponding increase in the share of the other regions. Notwithstanding this, the developed countries still continued to consume more than 50% of the world production except in the case of nickel where, in 1982, it reached only 46%.

Greater dynamism has been recorded, primarily in the group of the other developing countries where growing rates of minerals consumption have been recorded. Although both Latin America and the centrally-planned economy countries have increased their share of total world consumption, their annual rates have followed the same trend as that of the first group but this has been due rather to a greater decline in consumption in the group of developed countries.

Chapter II

THE EVOLUTION OF THE LATIN AMERICAN MINING SECTOR

1. Contribution to the formation of the gross domestic product

There is no definite pattern in the evolution of the contribution of the mining sector to the formation of the gross domestic product (GDP) in the different countries of Latin America, during the period 1960-1974 to 1974-1982. Whereas the growth rate of the GDP of extractive mining activities of a number of countries increased between both periods, in other cases this rate fell and even declined to negative values. Similar situations occurred during the subperiod 1980-1982 compared with the period 1974-1982. However, it must be borne in mind that the annual growth rate of the mining GDP for Latin America as a whole increased by 3.0% from the period 1960-1974 to 5.4% in the period 1974-1982, falling to 4.3% during the subperiod 1980-1982 (see table 3).

According to the evolution described, mining increased its share of the total GDP in one group of eight countries, whereas in another group of weight countries it declined and in yet a third group of three countries it remained constant (see table 4).

It is obvious that the explanation of the economic phenomena presumes a perception of the aspects which created it and clarify the why and the wherefore of their occurrence. For this, it would be necessary to have both a theoretical framework for the corresponding empirical research in each country of the region in order to determine the causes of the erratic behaviour observed in the evolution of the mining GDP. In the context of the theories of the development of the classical and neoclassical schools, based on production functions, natural resources constitute one of the strategic factors of that process. When these models were actually applied it was difficult to measure adeqluately the contribution made by this wealth. Subsequently, in analysis the Keynesian theory focused attention on the evolution of the components of the final demand and the accumulation of capital

Table 3

LATIN AMERICA: RELATIVE EVOLUTION OF MINING GROSS DOMESTIC PRODUCT
(Annual growth rates) a/
(Percentages)

Period 1960-1974		Period 1974-1982		Subperiod 1980-1982	
I. High growth rates		**I. High growth rates**		**I. High growth rates**	
1. Ecuador	24.02	1. Guatemala	23.03	1. Panama	15.00
2. Dominican Republic	15.18	2. Paraguay	22.71	2. Mexico	12.22
3. Brazil	11.09	3. Mexico	11.58	3. Paraguay	8.30
4. Honduras	9.92	4. Peru	6.17	4. Chile	6.82
5. Paraguay	8.26	5. Panama	5.56		
6. Panama	8.16	6. L. America's mining GDP	5.38		
7. Argentina	7.27	7. Uruguay	5.22		
8. Bolivia	7.26				
9. Mexico	6.52				
10. L. America's total GDP	5.93				
II. Lower growth rates		**II. Lower growth rates**		**II. Lower growth rates**	
1. El Salvador	4.58	1. Brazil	4.46	1. Colombia	4.95
2. Chile	3.27	2. L. America's total GDP	4.03	2. L. America's mining GDP	4.31
3. L. America's mining GDP	2.96	3. Chile	3.47	3. Brazil	4.16
4. Peru	2.93	4. Argentina	2.56	4. Peru	1.60
5. Nicaragua	1.90	5. Ecuador	1.70	5. Ecuador	1.45
6. Uruguay	1.86	6. Colombia	0.20	6. L. America's total GDP	0.25
7. Guatemala	0.37			7. Honduras	0.00
8. Venezuela	0.30				
9. Colombia	0.14				
III. Negative growth rates		**III. Negative growth rates**		**III. Negative growth rates**	
1. Haiti	-4.90	1. Honduras	-1.46	1. Argentina	-0.31
		2. Dominican Republic	-1.69	2. El Salvador	-2.67
		3. Bolivia	-3.67	3. Bolivia	-3.98
		4. El Salvador	-3.67	4. Haiti	-4.39
		5. Haiti	-5.00	5. Venezuela	-6.10
		6. Venezuela	-5.03	6. Nicaragua	-7.33
		7. Nicaragua	-17.43	7. Uruguay	-11.50
				8. Dominican Republic	-12.26
				9. Guatemala	-15.61

Source: ECLAC, Statistical Yearbook for Latin America - 1983.
a/ Constant 1970 prices.

LATIN AMERICA: SHARE OF THE MINING GROSS DOMESTIC PRODUCT IN THE TOTAL GROSS DOMESTIC PRODUCT a/

(In percentages) b/

Period 1960-1974	Period 1974-1982	Subperiod 1980-1982

I. Countries with a growing share

Period 1960-1974	Period 1974-1982	Subperiod 1980-1982
1. Bolivia 6.7 to 8.4	1. Chile 11.5 to 13.1	1. Chile 10.8 to 13.1
2. Ecuador 0.8 to 7.1	2. Peru 6.5 to 7.8	2. Bolivia 5.3 to 5.5
3. Dominican Republic 1.9 to 5.7	3. Mexico 2.4 to 3.6	3. Mexico 3.1 to 3.6
4. Honduras 1.6 to 3.2	4. Latin America 2.8 to 3.0	4. Latin America 2.8 to 3.0
5. Argentina 1.4 to 1.7	5. Argentina 2.2 to 2.7	5. Argentina 2.4 to 2.7
6. Uruguay 1.6 to 1.7	6. Uruguay 1.7 to 2.1	6. Brazil 0.7 to 0.8
7. Brazil 0.5 to 0.8	7. Paraguay 0.3 to 0.7	7. Paraguay 0.6 to 0.7
8. Paraguay 0.2 to 0.3	8. Guatemala 0.1 to 0.4	8. El Salvador 0.1 to 0.2

II. Countries with a constant share

Period 1960-1974	Period 1974-1982	Subperiod 1980-1982
1. Mexico 2.4	1. Brazil 0.8	1. Colombia 1.0
2. El Salvador 0.2	2. El Salvador 0.2	2. Honduras 2.1
3. Panama 0.2	3. Panama 0.2	3. Panama 0.2

III. Countries with a decreasing share

Period 1960-1974	Period 1974-1982	Subperiod 1980-1982
1. Venezuela 24.8 to 12.9	1. Venezuela 12.9 to 6.7	1. Peru 7.9 to 7.8
2. Chile 11.7 to 11.5	2. Bolivia 8.4 to 5.5	2. Venezuela 7.6 to 6.7
3. Peru 7.6 to 6.5	3. Ecuador 7.1 to 4.9	3. Ecuador 5.1 to 4.9
4. Latin America 4.2 to 2.8	4. Dominican Republic 5.7 to 3.5	4. Dominican Republic 4.8 to 3.5
5. Haiti 5.0 to 1.9	5. Honduras 3.2 to 2.1	5. Uruguay 2.4 to 2.1
6. Colombia 3.0 to 1.4	6. Colombia 1.4 to 1.0	6. Haiti 1.1 to 1.0
7. Nicaragua 1.1 to 0.6	7. Haiti 1.9 to 1.0	7. Guatemala 0.5 to 0.4
8. Guatemala 0.2 to 0.1	8. Nicaragua 0.6 to 0.1	8. Nicaragua 0.2 to 0.1

Source: See table 3 of the text.
a/ Mining GDP includes extractive activities of mining, quarries and hydrocarbons.
b/ Series in national currency at constant 1970 prices.

and natural resources lost the characteristic of being strategic variable. More recent theories are once agai conceding greater relative importance to the role of natura resources in the development process, for example, in givir to mining exploration a strategic value in the generation c foreign exchange (Perloff and Dodds in 1963); in the increasi in the public sector income and as the primary accumulation c an exhaustible source of wealth which would have to k converted into other forms of reproductive capital (Solow an Schulze in 1974, Pearce and Rose in 1975).

It would no doubt be useful to specify whether this se of basic principles would be the right and proper instrument for analysing the causes of fluctuations in minera production, the primary allocation of resources, th distribution of income generated by mining activity and th reallocation of the economic surplus. Thus for example, function of Cobb-Douglas on the production model 1/ at th aggregate level or at the level of mining enterprises complemented by the analysis of the opportunity cost o foreign exchange and the criterion of the reallocation of th economic surplus to exploration and mining research projects to mining infrastructure which will reduce the costs of minine production or other forms of reproductive capital in othe sectors, would be a technical framework that would have to be researched empirically in each country in order to demonstrate its validity as a functional analysis and programmine instrument.

2. Latin America's share of world mineral resources

Mining wealth is subject to ongoing evaluation depending oi how much is known about the size of the deposits and thei economic value which in turn hinges directly on the international contribution of minerals and metals in inverse relation to the extraction, production and marketing costs. I should be borne in mind that these resources are composed oi primary mineral which is found in earth and marine deposits and of secondary metal which can be obtained from the goods already used or which are not being exploited (scrap). The difficulties in interpreting and evaluating the information oi mining resources and the need to have common classificatior criteria prompted the United Nations Economic and Socia Council to approve a proposal, in March 1979, or "International Classification of Mining Resources" to enable their classification as follows:

R1: Proven reserves in situ, the details of which have been ascertained through prospecting and mining exploration work.
R2: Probable reserves, determined in a preliminary manner or inferred through continuity of known mineral veins.

R3: Potential reserves, known superficially through general geological prospecting work, continuity of minerological veins and formations or through surface exposure.

R1E: Proven reserves, which are potentially explorable in view of the price increase or cost reductions.

R1s: Reserves that are proven but which are economically marginal.

r1: Secondary metal.

The inventory of proven reserves (R1) in Latin America in 1981 showed that considerable wealth existed in both metallic and non-metallic minerals (see table 5). Compared with the production patterns of that year, there should be relatively larger deposits of potassium, uranium, iron ore, niobium, phosphated rocks, vanadium, nickel and lithium. If it is considered that mining projects have maturity periods of around ten years and that investment is justified if the reserves can guarantee that the project will last at least 20 years, it could be estimated that the region would have critical levels of reserves (lower than 20 years) for the following minerals: bismuth, silver, barite, zinc and gold.

The largest mineral reserves in Latin America compared with world reserves for that year would have the following percentages: niobium 82%, columbium 77%, lithium 59%, iron ore 45%, molybdenum 34%, copper 33%, bauxite 26%, selenium 26%, and bismuth 25%. From another viewpoint, the region would have a higher ratio of reserves to production compared with the other regions of the world, in the following group of minerals: iron ore, copper, nickel, uranium, asbestos and potassium. In contrast, this ratio would not be as favourable as those in other regions, in respect of zinc, tungsten, cobalt and chromium (see table 6).

A common denominator of the countries of the region is the need to have a better knowledge of mining resources. It is believed, for example, that exploration work has been done in only 5% of the potential mining area of Mexico and 10% of Bolivia. Furthermore, systematic work of the type done in Brazil, has enabled that country to become the principal mining producer of the region. The preliminary inventory of potential resources (R3) establishes that there may be large deposits of copper in Chile, tin in Brazil, manganese in Brazil and Bolivia, nickel in Cuba, silver and lead in Mexico.

In recent years, the majority of the countries of the region have begun new prospecting programmes, are concluding the preparation of their geological maps and have started the identification and possible tracking of mining deposits in preparing their respective metal-bearing maps. An analysis of these maps leads to the following preliminary conclusions:

a) Mexico has enormous mining potential which would have to be determined through semi-detailed geophysical

Table 5

THE MINING RESERVES OF LATIN AMERICA IN 1981

(In units of metric tons (MT))

Unit	Minerals	Totals	Argentina	Bolivia	Brazil	Chile	Colombia	Cuba	Mexico	Peru	Other countries
Thousands	Antimony	650		366					220	64	
Thousands	Asbestos	5 540			4 555		370		246		369
Thousands	Sulphur	90 000							90 000		
Thousands	Barite	15 378							9 087	6 291	
Millions	Bauxite	6 131			2 504						3 627
Thousands	Bentonite	1 270							1 270		
MT	Bismuth	23 655		13 585					5 510	4 560	
Thousands	Boron	36 288	7 856	10 288		9 144				9 000	
MT	Cadmium	69 440			19 800				19 720	29 920	
Millions	Carbon	40 000					38 000		1 000		1 000
MT	Cobalt	44 000			29 320						14 680
Thousands	Copper	189 445	5 000		625	107 069			25 000	34 518	17 233
MT	Columbium	8 165			8 165						
MT	Chromium	7 000			7 000						
Thousands	Quartz	4 258							4 258		
Thousands	Diatomite	1 180							1 180		
Thousands	Tin	1 587	10	1 000	406				10		161
Thousands	Strontium	365							365		
Thousands	Feldspar	3 020							3 020		
Thousands	Fluorite	52 419	10 000		3 300				39 087		32
Thousands	Graphite	1 418							1 418		
Thousands	Ilmenite	800			800						
Thousands	Yttrium	2			2						
Millions	Iron ore	53 773	120	26 066	20 827	360	240	3 500	360	600	1 700
Thousands	Lithium	1 299			9	1 290					
Millions	Magnesium	473			473						
Thousands	Manganese	73 000	30 795		40 370				1 835		

Table 5 (concl.)

Unit	Minerals	Totals	Argentina	Bolivia	Brazil	Chile	Colombia	Cuba	Mexico	Peru	Other countries
Thousands	Mercury	8 584	-	-	-	-	-	-	8 584	-	-
Thousands	Molybdenum	3 223	104	-	-	2 446	-	-	190	228	255
Thousands	Niobium	6 543	-	-	6 543	-	-	-	-	-	-
Thousands	Nickel	23 879	-	-	2 624	-	908	17 723	-	-	2 624
MT	Gold	387	-	-	-	-	-	-	387	-	-
MT	Silver	53 055	651	2 603	-	2 386	-	-	24 945	17 787	4 683
MT	Platinum	31	-	-	-	-	31	-	-	-	-
Thousands	Lead	13 163	940	627	2 507	-	-	-	5 014	4 075	0
Thousands	Potassium	64 000	-	-	54 480	9 080	-	-	-	-	440
Thousands	Rhenium	1 360	-	-	-	1 179	-	-	-	181	-
Millions	Phosphated rocks	1 490	-	-	922	-	-	-	568	-	-
Thousands	Rutile	58 000	-	-	57 800	-	-	-	-	-	200
MT	Selenium	57 264	-	-	-	34 044	-	-	5 423	12 798	4 990
MT	Tantalum	3 625	461	-	3 164	-	-	-	-	-	-
MT	Tellurium	3 200	-	-	-	-	-	-	-	3 200	-
Thousands	Fuller's earth	320	-	-	-	-	-	-	320	-	-
Thousands	Rare earths	318	-	-	318	-	-	-	-	-	-
Thousands	Thorium	54	-	-	-	-	-	-	-	-	-
Thousands	Tungsten	111	3	40	42	-	-	-	21	5	-
Thousands	Uranium	225	30	-	78	-	-	-	117	-	-
Thousands	Vanadium	223	-	-	-	127	-	-	-	-	96
Thousands	Gypsum	114 062	-	-	-	-	-	-	114 062	-	-
Thousands	Iodine	363	-	-	-	363	-	-	-	-	-
Thousands	Zinc	15 907	482	964	4 580	-	-	-	2 892	6 989	-
Thousands	Zirconium	895	-	-	893	-	-	-	2	-	-

Source: See table 25 of the Statistical Appendix.

41

Table 6

LATIN AMERICA: SHARE OF WORLD RESERVES, PRODUCTION AND CONSUMPTION OF MINERALS - 1980/1981

Minerals	Participation in percentages				Ratios of reserves to:		Ratios of reserves to:		Production of other regions	
	Reserves	Production minerals	Production metals	Consumption	Consumption	Production	Other developing countries	Developed countries	Centrally planned economy countries	World total
Niobium	82	82.2	...	...	...	503	...	...	...	...
Columbium	77	...	...	...	...	...	...	...	...	...
Lithium	59	59.0	1.2	1.2	16 443	332	...	...	...	...
Iron ore	45	18.3	3.5	3.1	3 360	577	7	172	180	233
Molybdenum	34	13.4	1.0	2.8	1 074	230	904	60	76	87
Copper	33	20.6	12.6	6.4	381	117	77	60	39	73
Bauxite	26	27.2	5.7	3.9	1 686	243	544	156	68	252
Selenium	26	16.7	1.0	1.5	1 974	180	...	...	...	...
Bismuth	25	37.4	16.8	18.4	38	18	...	...	...	...
Nickel	24	9.0	5.3	2.1	1 492	356	218	55	120	134
Silver	23	32.2	...	2.9	176	15	...	...	...	...
Fluorite	17	19.7	-	19.8	56	56	80	111	17	64
Tin	16	15.4	12.8	4.7	144	44	34	42	70	42
Antimony	15	29.9	3.6	8.2	122	33	...	...	...	...
Cadmium	10	5.1	2.1	3.1	119	72	...	...	...	...
Uranium	9	0.4	0.4	0.8	639	1 203	33	59	...	58
Lead	8	10.3	8.3	7.8	47	35	30	56	26	43
Barite	7	15.5	-	16.5	12	13	...	...	...	...
Zinc	7	14.7	7.0	5.6	45	17	43	52	21	38

Table 6 (concl.)

Minerals	Participation in percentages				Ratios of reserves to:		Ratios of reserves to:		Production of other regions	
	Reserves	Production minerals	Production metals	Consumption	Consumption	Production	Other developing countries	Developed countries	Centrally planned economy countries	World total
Tantalum	6	...	...	...	...	-	...	...	...	...
Asbestos	5	2.9	-	6.7	16	39	21	33	17	25
Mercury	5	1.1	0.8	5.3	24	117	...	...	...	...
Tellurium	5	18.8	...	0.2	3 200	34	...	...	...	...
Manganese	8	10.6	10.6	11.2	50	53	8	123	58	68
Tungsten	4	10.2	3.4	3.4	60	20	29	40	64	49
Phosphated rocks	2	2.2	-	4.7	235	493	3 316	203	290	521
Zirconium	2	0.6	...	...	...	...	...	...	...	...
Cobalt	2	5.3	...	1.5	89	25	98	83	265	112
Gold	1	6.5	1.9	2.2	15	4	...	...	...	...
Potassium	1	0.1	-	6.3	36	2 286	511	235	430	325
Vanadium	1	1.3	1.3	1.7	384	490	108	416	536	469
Chromium	less than 1	3.6	3.3	4.3	17	20	478	628	58	363
Magnesium	less than 1	-	-	6.2	24	-	...	...	...	...
Platinum	less than 1	0.2	-	23.6	less than 1	77	...	275	62	173
Rutile	less than 1	0.1	-	...	1	129	153	39	193	64

Source: 1. Mining production, reserves and consumption: see tables 3, 6 and 9 of the Statistical Appendix.
2. Metallurgical production: Metallgesellschaft Aktiengesellschaft, Metal Statistics, 1972-1982.

or geochemical exploration in an area estimated at over 1.5 million square kilometers.

b) The majority of the countries of Central America have a geological structure with metal-bearing features similar to the volcanic formations of the Sierra Madre in Mexico and the potential has not been duly explored. The copper-bearing district of Panama may also extend to other countries of this subregion.

c) The respective maps of the Andes mountain range identify a vast mining potential which could be extensively and intensively explored.

d) The territory lying between the outlets of the Orinoco and Amazon rivers could become another metal-bearing province of prime importance once the appropriate procedures have been laid down to facilitate access to the interior of the jungle.

e) Brazil's size and excellent metal-bearing features are making it possible to utilize remote sensors intensively for example, in the Radam-Brazil project.

f) Argentina may also be able to increase its mining production if detailed exploration of the Andean region is concentrated in the area extending from the province of Jujuy to the Neuquén.

g) The eastern part of Paraguay has a geological structure which suggests that there are copper and carbon deposits which could be confirmed through field exploration.

3. The evolution of mining production

In order to arrive at a number of behaviour patterns of mining production, first of all, a complementary analysis was done worldwide, of the period prior to the world crisis, 1965-1974, the first period of the world crisis 1974-1980 and the period during which the mining crisis worsened, 1980-1983 (see table 7).

The production of the majority of the 35 products analysed achieved moderate annual growth rates which fluctuated between 0.4% (tantalum) and 4.7% (chromium) during the first period. This group includes the following major export products of Latin America: copper, tin, iron ore, silver, lead and zinc. During this period, about one third of the minerals reached annual growth rates of world production that fluctuated between 5% (selenium) and 18.5% (phosphated rocks). This group includes the following major export minerals of Latin America: bauxite, nickel and tungsten. On the other hand, only four minerals showed negative expansion average rates which ranged between -0.3% (mercury) and -4.1% (cadmium).

During the period 1974-1980 there was a group of products with annual negative expansion rates ranging from -0.1% (zinc) to -9.1% (iron ore). Except for a group of seven minerals, the

ANNUAL GROWTH RATES OF THE WORLD PRODUCTION OF THE MAIN MINERALS

Minerals	Period 1965-1974			Period 1974-1980			Period 1980-1983		
	High rates	Low rates	Negative rates	High rates	Low rates	Negative rates	High rates	Low rates	Negative rates
Antimony		1.5				-1.9			-9.1
Metallic arsenic		2.8	-2.1						-5.2
Asbestos		4.4			1.1				-5.3
Sulphur		3.0							-1.8
Barite	8.6				3.2				-9.7
Bauxite		1.4				-5.6	6.7		
Bismuth						-0.3			
Cadmium	8.6				0.1				
Cobalt		4.1				-0.5		1.7	
Copper	8.5								
Columbium									-6.8
Chromium						-0.6			-3.7
Tin		1.8				-9.1			-2.8
Fluorite	6.0				0.6		13.8		
Iron ore		4.3				-2.5			-6.3
Ilmenite		1.4							
Lithium	16.0			16.0			138.1		
Magnesium		3.5			2.7				-5.1
Manganese		2.9				-0.9			-16.4
Mercury			-0.3		3.9	-5.3			-4.3
Molybdenum	7.1					-2.2			
Nickel		3.0			0.4				
Gold	7.7				1.1		5.5		
Silver	6.7				0.9		5.9		-1.6
Platinum		1.8	-1.9						
Lead		2.9				-1.3			-1.4
Potassium	18.5			5.1					
Phosphated rocks		4.6			1.8				
Rutile	5.0			7.6					
Metallic Selenium		0.4		16.5					
Tantalum		3.2		6.2					
Metallic tellurium	9.0								
Tungsten	10.1								-8.2
Vanadium								1.3	
Zinc		3.1				-0.1			-0.1

Source: See table 1 of the Statistical Appendix.

expansion rates of the others had fallen lower than the rates achieved during the earlier period.

The last period 1980-1983, showed a predominance of one group of 15 minerals with negative expansion rates ranging from -1.4% (lead) to -16.4% (molybdenum). However, it should be mentioned that there was notable recovery, during this period, in iron ore, gold, silver, lithium and bismuth production.

It can be seen from the analysis of 20 products, that Latin American mining production expanded more than world production of 14 minerals, during the period 1960-1980. This group includes the following main export products: copper, tin, iron ore, nickel and tungsten. Nevertheless, the highest growth rates were achieved by "non-traditional minerals" such as vanadium, cobalt, platinum, molybdenum and uranium.

The greatest differences were noted during the period 1980-1983, when the growth rate of world production increased in only one product: iron ore, whereas Latin American production achieved growth in the following eight minerals: copper, tin, iron ore, molybdenum, platinum, lead, tungsten and zinc. At the other end of the scale, world production showed negative rates for 12 minerals whereas in Latin America, that same group was composed of seven minerals but with greater reductions than the world level in the case of asbestos (-1.0%), bauxite (-11.1%), chromium (-23.0%), cobalt (-59.1%) fluorite (-9.2%), manganese (-6.3%) and nickel (-7.8%) (see table 8).

4. Consumption levels of the major minerals

As indicated before, the impact of the crisis on levels of mining production would be linked to the effect of the crisis on the internal levels of consumption and international demand. Looking at the group of major metallic minerals exported from Latin America it can be observed that during the period 1950-1960, the annual growth rate of consumption in the region fluctuated between 2.4% for tin and 11.5% for bauxite. During the period 1960-1980, the variation was 3.3% for iron ore and 15.1% for nickel. For iron, the growth rates of its consumption were 7.4% in the subperiod 1965-1974, -2.8% in the subperiod 1974-1980 and 19.9% in the subperiod 1980-1983 (see table 9).

During the period 1960-1980, the annual growth rates of consumption in this group of minerals in almost all cases doubled the rates for mining production of the region, showing extreme proportions between 50% for iron ore and 277% for zinc. Despite these high consumption rates, the proportion of production earmarked, in 1984, for regional consumption fluctuated between 15% for bauxite and 76% for lead. However, if the difference indicated between the growth rates of production and consumption is maintained it can be assumed

ANNUAL GROWTH RATES OF THE RELATIVE EVOLUTION OF MINERAL PRODUCTION

Minerals	Latin America		Other developing countries		Developed countries		Centrally planned economy countries		World total	
	1960-1980	1980-1983	1960-1980	1980-1983	1960-1980	1980-1983	1960-1980	1980-1983	1960-1980	1980-1983
Asbestos	6.1	-1.0	5.3	-10.0	1.9	-10.9	6.5	-0.6	4.1	-5.2
Bauxite	3.1	-11.1	10.1	-4.2	9.2	-4.0	4.2	-0.6	6.2	-5.3
Chromium	5.3	-23.0	0.5	-8.2	6.3	-10.2	5.4	-1.6	4.0	-6.8
Cobalt	18.1	-59.1	3.6	-11.7	5.2	-1.2	6.5	2.1	4.3	-9.7
Copper	3.5	8.3	2.7	-0.6	2.7	-1.9	5.4	3.5	3.4	1.7
Tin	2.4	5.4	1.5	-8.4	4.5	-1.9	-1.8	1.8	1.1	-3.7
Fluorite	4.7	-9.2	14.9	-6.6	3.3	-8.9	5.3	6.5	4.9	-2.8
Iron ore	6.5	19.4	4.7	1.3	2.3	5.3	4.6	23.8	3.8	13.8
Ilmenite	-	-	-0.02	-	5.4	-	14.4	-	5.0	-
Manganese	4.8	-11.8	2.2	-4.0	7.6	-17.5	2.4	0.2	3.4	-6.3
Molybdenum	10.6	18.5	9.7	-15.7	4.7	-32.8	3.3	4.4	5.0	-16.5
Nickel	7.2	-7.8	7.1	-6.6	2.2	-9.6	5.1	5.2	4.0	-4.3
Platinum	13.9	15.5	-24.0	217.5	7.0	-7.1	12.1	3.5	8.7	-1.7
Lead	0.3	8.3	0.2	-7.8	2.5	-2.1	3.0	-2.2	2.1	-1.4
Potassium	-5.3	-	12.1	-	4.7	-	7.3	-	5.8	-
Phosphated rocks	6.2	-	5.5	-	6.0	-	7.1	-	6.1	-
Rutile	4.0	-	20.5	-	6.6	-	28.5	-	7.8	-
Tungsten	4.4	5.2	2.7	-12.8	3.8	-14.7	1.9	-6.8	2.7	-8.6
Uranium	7.4	-	17.0	-	3.3	-	-	-	4.4	-
Vanadium	28.5	-	-	-	6.4	-	27.9	-	8.5	-
Zinc	2.8	4.3	1.2	3.3	3.1	1.1	3.9	-0.4	3.2	1.3

Source: See table 3 of the Statistical Appendix.

47

Table 9

EVOLUTION OF THE WORLD CONSUMPTION OF THE MAIN MINERALS

(Annual growth rates)

(Percentages)

Minerals	Period 1965-1974 High rates	Period 1965-1974 Low rates	Period 1974-1980 High rates	Period 1974-1980 Low rates	Period 1974-1980 Negative rates	Period 1980-1983 High rates	Period 1980-1983 Low rates	Period 1980-1983 Negative rates
I Bauxite								
a) Latin America	...	...	7.2	-	-	-	-	-10.9
b) Other developing countries	...	...	10.4	-	-	-	2.3	-
c) Developed countries	...	...	-	0.4	-	-	-	-4.4
d) Centrally planned economy countries	...	...	-	3.9	-	-	-	-4.3 a/
TOTAL	...	...	-	1.8	-	-	-	-8.5 a/
II Cadmium								
a) Latin America	22.5	-	13.3	-	-	7.6	-	-
b) Other developing countries	9.8	-	16.0	-	-	9.7	-	-
c) Developed countries	-	3.0	-	3.2	-	-	-	-0.8
d) Centrally planned economy countries	5.4	-	-	0.0	-	-	-	-2.2 a/
TOTAL	-	3.6	-	-	-1.6	-	-	-1.5 a/
III Copper								
a) Latin America	7.3	-	7.9	-	-	-	-	-13.3
b) Other developing countries	7.3	-	12.5	-	-	11.4	-	-
c) Developed countries	-	2.6	-	0.4	-	-	-	-1.8
d) Centrally planned economy countries	5.2	-	-	4.6	-	-	0.5 a/	-
TOTAL	-	3.4	-	2.1	-	-	1.9 a/	-
IV Tin								
a) Latin America	4.6	-	-	3.4	-	-	-	-3.1
b) Other developing countries	-	0.0	-	2.8	-	-	2.5	-
c) Developed countries	-	1.3	-	-	-3.1	-	-	-2.9
d) Centrally planned economy countries	3.0	-	-	1.2	-	-	-	-0.9 a/
TOTAL	-	1.7	-	-	-1.6	-	-	-5.1 a/

Table 9 (concl.)

Minerals	Period 1965-1974			Period 1974-1980			Period 1980-1983		
	High rates	Low rates	Negative rates	High rates	Low rates	Negative rates	High rates	Low rates	Negative rates
V Iron ore									
a) Latin America	7.4	-	-	10.3	-	-	19.9 a/	-	-
b) Other developing countries	-	1.6	-	-	3.5	-	8.9 a/	-	-
c) Developed countries	5.1	-	-	-	-	-2.8	-	-	-12.2 a/
d) Centrally planned economy countries	-	4.2	-	-	-	-2.0	-	-	-2.0 a/
TOTAL	-	4.7	-	-	0.4	-	-	-	-5.8 a/
VI Magnesium									
a) Latin America	25.1	-	-	26.0	-	-	11.8 a/	-	-
b) Other developing countries	-	-	-	-	4.9	-	-	-	-29.3 a/
c) Developed countries	5.1	-	-	-	-	-2.2	-	-	-8.1 a/
d) Centrally planned economy countries	7.0	-	-	5.2	-	-	-	1.2 a/	-
TOTAL	6.0	-	-	-	0.2	-	-	-	-6.4 a/
VII Nickel									
a) Latin America	20.9	-	-	6.4	-	-	8.1	-	-
b) Other developing countries	10.7	-	-	11.3	-	-	-	0.0	-
c) Developed countries	5.8	-	-	-	3.2	-	-	-	-5.8
d) Centrally planned economy countries	-	-	-	-	-	-1.2	-	2.1	-
TOTAL	5.6	-	-	-	0.2	-	-	-	-6.4
VIII Lead									
a) Latin America	6.9	-	-	7.7	-	-	6.3	-	-
b) Other developing countries	9.5	-	-	-	1.7	-	-	-	-8.1
c) Developed countries	6.3	-	-	-	3.6	-	-	0.0 a/	-
d) Centrally planned economy countries	5.2	-	-	-	-	-0.3	-	-	-1.2
TOTAL	-	4.5	-	-	1.2	-	-	-	-1.1 a/
IX Zinc									
a) Latin America	8.7	-	-	6.5	-	-	6.1	-	-
b) Other developing countries	8.5	-	-	7.9	-	-	-	-	-4.9
c) Developed countries	-	2.8	-	-	2.3	-	-	1.4 a/	-
d) Centrally planned economy countries	8.0	-	-	-	-	-1.5	-	-	-0.1
TOTAL	-	4.3	-	-	0.5	-	-	-	-1.7 a/

Source: See table 7 of the Statistical Appendix.
a/ Period 1980-1982.

49

that by the year 2000, the greater part of minerals production in Latin America would be consumed within the region.

Considering that minerals consumption is to a large extent dependent on industrial expansion, a more detailed analysis of this consumption would have to include the technical ratios existing between both the industrial and mining sectors in respect of mining output and by countries of the region. Between 1960 and 1974, the annual industrial growth rate of Latin America was approximately 7.0%, whereas the corresponding rate for mining production reached 3.3%. On the whole, a relatively close correlation was established ($R2=0.78$) between both growth rates for this period. During period 1974-1980, the correlation between both sectors was even closer ($R2=0.94$), but conversely since while the industrial sector showed a decline in the rate of expansion, the mining sector increased its growth rate, perhaps because of its improved bargaining power on the international market or because of the greater expansion of that market. This situation could be showing that the surplus of mining production has found an appropriate outlet through exports.

During the period 1980-1982, whereas the growth rates of industrial production maintained their downward trend, the mining sector showed a sharp decline in 1981 (going from 12.8% to 0.6%) and a significant recovery in 1982 (11.8%); as a result, during this period, the ratio also fell ($R2=0.53$). The drop in growth rates of industrial production was caused primarily by the decline in production in Argentina, Brazil, Peru and Venezuela, in 1981 and in Argentina, Colombia, Chile, Mexico and Peru, in 1982. At the same time, the decline in manual production in 1981 was caused primarily by the drop in production in Brazil, Colombia, Ecuador, Guyana, Honduras, Jamaica, Nicaragua, Peru, the Dominican Republic and Suriname. The positive rate achieved by mining production in the region in 1982 was caused primarily by the recovery of production levels in Brazil (6.4%) and the high production rates in Chile (15.1%), since it must be borne in mind that during that year the mining production rates were negative in Argentina, Bolivia, Cuba, Ecuador, Guyana, Haiti, Jamaica, the Dominican Republic, Suriname and Venezuela.

With respect to the main export products from Latin America, the following ratios have been established between the impact of the prices on consumption levels and of these levels on production levels (see table 9).

a) <u>Bauxite</u>: During the period 1970-1980, whereas consumption in Latin America grew at an annual rate of 7.2%, production fell at the rate of -2.7%. During the following period, 1980-1983, production continued to decline but at an annual rate of -0.3% following a decline in consumption at annual rates of -10.9%. This difference in the performance of the consumption and production of bauxite caused not only a low correlation coefficient of ($R2=0.12$) but also a low inverse coefficient, which showed that production was more dependent on the vagaries of foreign trade and other diverse factors.

b) Copper: The annual growth rates of copper consumption in the region were 7.3%, in 1965-1974, of 9% in 1974-1980 and -13.3% in 1980-1983. For their part, the corresponding production rates were 3.8%, 2.8% and -3.6% and in this case had a high correlation coefficient of (R^2=0.86) for the annual data over the period 1965-1983, which would mean that for each 1% of growth in copper production, consumption would have to increase at rates of 2% to 4%.

c) Tin: The annual growth rates of regional tin consumption were 4.1% and -3.1% for the periods 1965-1980 and 1980-1983 respectively. The corresponding regional production rates were 2.0% and 0.9% between both periods. As the correlation coefficient between consumption and production is relatively high (R^2=0.64), for every 2% increase in consumption, production could be expected to increase at around 1%.

d) Iron ore: During the period 1965-1974, the growth rates of regional consumption and production of iron ore were 7.4% and 11.7% respectively. During the subsequent period 1974-1980, these rates fell to -2.8% for consumption and -4.7% for production, rising in the following period 1980-1982 to 19.9% for consumption and 8.2% for production. With a relatively high correlation coefficient between both variables (R^2=0.64) it is fairly certain, that for each increase in consumption of 2%, production will increase at around 1%.

e) Nickel: The annual growth rates of nickel consumption and production in the region showed a declining trend since consumption rates fell from 20.9% during the period 1965-1974 to zero during the period 1980-1983 and production rates fell from 11.3% during the first period to -4.1% during the second period. With a high correlation coefficient (R^2=0.87) it can be reliably expected that for each increase in consumption between 2% and 4%, production would grow at around 1%.

f) Lead: The growth rates of regional lead consumption also showed an upward trend, with annual values of 6.9% during the period 1965-1974, 1.7% during the period 1974-1980 and 1% during the period 1980-1982. On the other hand, production rates in those periods were as follows: 2.5%, -4.6% and 9.8%, with the historical series of annual data showing a very low production to consumption ratio (R^2=0.17).

g) Zinc: Regional consumption and production of zinc both showed an upward trend in their expansion rates, with negative figures during the last period. The high correlation coefficient (R^2=0.92) means that for every 1% increase in production a 3% increase in consumption would almost certainly be needed.

Another of the basic features of the ratio between the mining and industrial sectors is in fact the low level of industrial profit earned from mining production earmarked specifically for the international market. Among the metallic minerals which had the lowest proportions processed industrially, in 1980, were lithium, selenium, molybdenum, antimony and iron ore with proportions processed of 2% to 19%. On other hand, the products with the highest proportions processed of between 50% to 100%, were vanadium, manganese, chromium, tin, lead and mercury. The growing demands of regional consumption and the need to substitute smaller volumes of exports for higher added value would determine how much effort would be needed to increase metallurgical production in the region and would perhaps require greater volumes than would have been destined for mining production. In this sense, in order to make the best use of the technical and economic advantages of economies of scale and of the integrated plants and the concentration of the continent's mining reserves, an analysis must be made to determine whether a single regional metallurgical structure or at least highly complementary industrial structures can be established in the different countries, depending on the availability of minerals and the opportunities for industrial development.

Minerals consumption, per capita, in Latin America is still far below that of the group of developed countries. It ranges from 2% to 4% of consumption per capita in these countries for uranium, selenium, cobalt and gold and in proportions of 73% to 84% for bauxite, barite, bismuth and manganese (see table 10).

Perhaps its metallic infrastructure requirements, its need to reconvert its thermal energy sources using hydroelectric energy and to increase both its domestic and foreign means of transport and the machinery and metallic equipment requirements of the mining metallurgical sector itself and of other industrial agricultural and service sectors are keeping Latin America's mineral and metal consumption at the growth rates of past decades. This situation will continue, since the annual population growth rates will be around 2.5% and the income elasticity of these products went from 1.8% to 3.1% during the period 1963-1973.

5. Latin American trade

In 1980, Latin America's minerals and metals exports consisted primarily of 19 products, nine of which represented about 95% of the value of those exports: copper (36%), iron ore (23%), zinc (11%), bauxite (8%), silver (7%), gold (4%), nickel and tin (3% each) and lead (1%). Antimony, lithium, molybdenum, bismuth, cadmium, cobalt, rutile, selenium, tellurium and tungsten all had lower percentages.

During the period 1970-1977, the highest growth rates of Latin America's minerals and metals exports in terms of volume, were in the metals exports of tin, raw steel, zinc, lead and aluminium with corresponding reductions in the exports of their respective minerals and concentrates. The

Table 10

ESTIMATE OF CONSUMPTION PER CAPITA OF MINERALS - 1980

Unit	Minerals	Latin America	Argentina	Brazil	Mexico	Peru	Venezuela	The other countries	Other developing countries	Developed countries	Centrally planned economy countries
Gr	Antimony	15.07	25.56	18.05	33.13	1.67	1.88	0.30	5.04	0.35	1.06
Kg	Asbestos	0.93	0.67	1.46	0.86	0.44	0.63	0.54	7.75	0.37	0.58
Kg	Barite	3.54	2.41	0.93	4.77	3.94	3.38	6.19	4.37	0.82	3.37
Kg	Bauxite	10.30	12.81	14.47	9.61		36.13	2.63	3.75	0.13	0.73
Gr	Bismuth	1.78	0.85	0.21	8.23			0.03	4.14	0.84	6.36
Gr	Cadmium	1.63		1.43	5.71				6.04	0.10	0.63
Kg	Chromium	1.18	0.19	2.85	0.73	0.22	0.44	0.10	1.71	0.20	0.49
Gr	Cobalt	1.40	5.11	2.28	0.77		0.63	0.58	3.68	0.04	0.39
Kg	Copper	1.41	1.96	1.99	1.76	1.06		0.03	7.42	0.18	0.87
Kg	Tin	0.03	0.04	0.04	0.03			0.04	3.00	0.17	0.75
Kg	Fluorite	2.63	0.07		13.11	0.11	0.06		8.22	1.30	2.41
Kg	Iron ore	45.33	50.22	75.37	52.97			16.82	2.86	0.13	0.34
Gr	Lithium	0.22	0.96	0.43					2.00	0.04	0.21
Kg	Magnesium	0.06		0.05	0.26					0.26	1.20
Kg	Manganese	8.45	5.78	17.91	7.03	0.11	0.13	1.28	3.81	0.84	1.04
Gr	Mercury	0.99	2.22	1.52	0.97		0.38	0.30	7.62	0.22	0.57
Kg	Molybdenum	0.01		0.02	0.01					0.08	1.00
Kg	Nickel	0.05		0.09	0.04			0.02	5.00	0.08	0.38
Gr	Gold	0.07	0.04	0.09		0.22	0.13	0.08	7.00	0.06	0.37
Gr	Silver	0.85	1.67	0.98			0.94	1.21	1.93	0.10	0.49
Gr	Platinum	0.14	0.30			2.28	0.06		7.00	1.08	2.00
Kg	Lead	0.79	1.70	0.67	1.37	1.44		0.29	5.64	0.19	0.74
Kg	Potassium	4.99	1.44	8.60	2.03	0.39	0.31	5.14	8.60	0.26	0.67
Kg	Phosphated rocks	17.94	1.85	33.01	23.16	0.78	1.19	5.75	3.78	0.16	0.81
Gr	Selenium	0.08	0.52		0.19			0.02	4.00	0.04	0.47
Gr	Tellurium		0.04								
Gr	Tungsten	5.19	2.22	9.97	4.16	13.89	0.06	0.04	3.55	0.18	0.29
Gr	Uranium	1.00	3.22	1.32				0.03	3.45	0.02	
Gr	Vanadium	1.65		4.53				0.24	4.71	0.07	0.18
Kg	Zinc	0.99	1.15	1.12	1.27	1.28	1.63	0.41	3.96	0.21	0.83

Latin America in proportion to the consumption per capita of other regions

Source: See tables 9 and 10 of the Statistical Appendix.

53

Table 11

EVOLUTION OF THE WORLD AND LATIN AMERICAN EXPORTS OF THE MAIN METALS AND MINERALS

(Annual growth rates of the trade volumes)

Exports of:	Europe		Japan		Main importers United States		Other countries		Totals	
	1970-1977	1977-1981	1970-1977	1977-1981	1970-1977	1977-1981	1970-1977	1977-1981	1970-1977	1977-1981
I Aluminium										
a) World	1.3	0.2	8.4	23.0	9.8	1.4	-10.8	166.7	3.8	5.1
b) Latin America	-8.8	7.8	...	70.9	98.2	21.8	-	-	-2.1	60.8
I A Bauxite										
a) World	9.6	0.9	5.5	-4.9	-1.2	1.1	7.0	-4.4	3.3	-0.8
b) Latin America	1.0	10.1	9.9	-10.6	-3.7	-2.2	-9.5	15.4	-4.2	0.6
II Metallic copper										
a) World	1.7	-3.7	-1.4	4.9	2.9	2.1	50.6	-4.2	1.6	-2.1
b) Latin America	0.9	1.1	4.5	9.7	-2.2	10.1	-	-	0.5	4.3
II A Copper concentrates										
a) World	13.2	-2.9	8.8	4.3	8.2	-11.8	-	-	9.6	2.5
b) Latin America	-1.1	21.6	5.3	13.5	-6.3	28.5	-	-	2.5	16.7
III Metallic tin										
a) World	-0.3	-1.2	0.7	1.7	-1.0	-1.0	1.3	5.4	-0.2	0.2
b) Latin America	...	5.6	...	...	46.5	13.2	47.7	-20.8	44.3	2.4
III A Tin concentrates										
a) World	-5.4	-15.8	-	-	5.1	-48.7	-	-	-3.7	-20.3
b) Latin America	-9.2	-24.2	-	-	5.0	...	-	-	-5.5	-23.9

Table 11 (concl.)

Exports of:	Europe		Japan		United States		Other countries		Totals	
	1970-1977	1977-1981	1970-1977	1977-1981	1970-1977	1977-1981	1970-1977	1977-1981	1970-1977	1977-1981
IV Metallic lead										
a) World	-0.6	-1.9	49.7	19.2	0.6	-18.2	2.3	-9.0	0.3	-4.7
b) Latin America	4.8	-19.6	...	42.1	3.1	-22.4	...	-22.9	4.6	-16.2
IV A Lead concentrates										
a) World	-6.8	3.9	-1.1	7.2	-5.9	-16.0			-5.4	2.5
b) Latin America	-15.0	13.0	3.6	10.0	-1.6	-35.8	-		-5.8	-1.4
V Metallic zinc										
a) World	8.7	-3.4	4.0	1.9	11.4	3.6	-4.7	-2.1	8.9	0.1
b) Latin America	16.5	-27.6	-5.4	-22.8	3.2	8.3	-		5.8	0.8
V A Zinc concentrates										
a) World	2.2	-0.9	-0.1	-2.3	-18.8	1.4			-0.1	-1.2
b) Latin America	18.2	-1.4	-1.6	-3.9	-19.6	8.2			3.6	-1.8
VI Raw steel a/										
a) World	1.2	-0.9	-		7.0	-5.2	8.4	-1.3	4.3	-1.3
b) Latin America	-		-		7.8	1.2	10.9	21.9	9.6	14.9
VI A Iron ore concentrates b/										
a) World	-0.6	-0.3	-4.5	2.5	-10.4	-5.6			-3.8	0.3
b) Latin America	0.4	5.9	-4.8	6.1	-21.3	-13.8			-6.5	3.4

Source: See table 14 of the Statistical Appendix.

a/ Period 1973-1978 and 1978-1982.

b/ Period 1975-1978 and 1978-1981.

exception to this group of products was copper, whose exports of both minerals and concentrates had higher growth rates than those of blister and refined copper. However, in both cases the growth rates were lower than those achieved in world production and this was also true for world exports of aluminium, bauxite and metallic zinc. Finally, from table 11 it will be seen that the greatest variations occurred in exports to the United States and European markets.

During the subsequent period 1977-1981, the same situation that had obtained in the previous period continued, except in the case of lead, whose exports of minerals and concentrates fell much less. During this period, exports of aluminium, copper concentrates, raw steel and iron ore increased. The greater dynamism of these products was due mainly to the wider fluctuations in exports to the markets of Japan (aluminium and iron ore), the United States (copper minerals) and other countries (raw steel).

In terms of value at constant 1975 prices, the annual growth rates of the exports of this group of minerals and metals, which was 2.8% during the period 1970-1974, fell at a rate of -0.7% per year during the period 1974-1980 and fell even further during the period 1980-1982, at an annual rate of -2.7%. During this latter period, positive rates were observed in all the exports from Brazil, except for exports of silver; copper, tin, iron-steel, silver and zinc from Peru; copper and steel from Mexico; zinc from Bolivia and bauxite-aluminium from Venezuela (see table 12). On the whole, the exports of these products grew during this period at annual rates of 8.8% in Brazil, 0.6% in Peru and 2.4% in Venezuela. The other countries showed negative growth rates ranging from -0.4% in Cuba to -48.7% in the Dominican Republic. As a result of this very uneven evolution, in 1982 the share of the different countries in the exports of minerals and metals from the region were as follows: Bolivia 4.6%, Brazil 27.4%, Chile 24.8%, Cuba 3.3%, Guyana 1.2%, Jamaica 6.8%, Mexico 4.0%, Peru 15.5%, the Dominican Republic 0.4%, Suriname 4.1% and Venezuela 7.9%. It should be borne in mind that the exports of two or three countries account for a high percentage of the region's exports of each product and that, therefore, fluctuations in the international market of particular products in some cases affect one group of countries and other products another group, which makes it difficult to organize large groups of regional producers per product (see table 13).

The figures in table 14 indicate that the region's share in world minerals and metals exports fell from 7.1% to 5.7% between 1970 and 1980 and increased to 6.0% in 1983. During the first period, the largest reductions were observed in the region's exports to Canada, the countries of the European Community, the centrally-planned economy countries, the United States and other developed countries but the share of interregional exports and exports to Japan and other developing countries increased. During the second period, the

Table 12

LATIN AMERICA: EVOLUTION OF THE EXPORTS OF THE MAIN MINERALS AND METALS

(Annual growth rates at constant 1975 prices)

(Percentages)

Products/Exporting countries	1970-1974	1974-1980	1980-1982
Bauxite-Alumina-Aluminium	1.7	5.0	-3.5
a) Brazil	30.9	94.7	23.5
b) Guyana	-7.4	1.0	-22.8
c) Jamaica	6.2	-3.3	-10.8
d) Mexico	-36.8	...	...
e) Dominican Republic	-9.8	-8.6	-44.8
f) Suriname	-4.0	14.7	-8.4
g) Venezuela	1.7	54.1	19.6
Copper	-0.4	-4.5	-6.8
a) Bolivia	-7.8	-29.4	-2.4
b) Brazil	5.8	13.2	45.9
c) Chile	1.2	-6.2	-8.0
d) Mexico	9.6	26.8	22.9
e) Peru	-6.9	-0.9	-11.2
Tin	7.7	-0.7	-6.7
a) Bolivia	6.2	-1.3	-11.0
b) Brazil	31.5	3.3	13.7
c) Mexico	...	...	...
d) Peru	32.8	19.5	48.5
Iron ore - Steel	3.4	3.9	5.1
a) Brazil	11.7	9.6	8.2
b) Chile	1.0	-6.5	3.9
c) Mexico	...	150.0	3.7
d) Peru	-15.5	-4.6	18.6
e) Venezuela	-2.1	-8.0	-19.9
Nickel	4.8	-0.4	-10.5
a) Brazil	...	...	...
b) Chile	...	...	-0.4
c) Cuba	-8.5	3.7	-0.4
d) Dominican Republic	...	-7.8	-49.4
Silver	15.5	-4.4	5.0
a) Bolivia	9.5	16.3	-41.9
b) Brazil	61.0	3.1	-78.1
c) Chile	1.6	37.3	-14.5
d) Mexico	21.4	-10.7	72.1
e) Peru	12.8	...	...
Lead	3.5	2.3	-26.1
a) Bolivia	-4.5	-5.7	-30.2
b) Brazil	...	...	...
c) Chile	...	...	...
d) Mexico	10.4	-11.0	-38.7
e) Peru	1.3	7.9	-23.7
Zinc	19.7	-9.9	8.4
a) Bolivia	10.4	-9.6	6.0
b) Brazil	...	...	...
c) Chile	...	...	...
d) Mexico	21.7	-12.7	-23.9
e) Peru	20.3	-7.8	22.8
Totals	2.8	-0.7	-2.7

Source: See table 16 of the Statistical Appendix.

Table 13

LATIN AMERICA: COMPOSITION OF SELECTED EXPORTS OF MINERALS AND METALS
BY MAIN EXPORTING COUNTRIES
(Percentages)

Products/Exporting countries	1970	1974	1980	1982
Bauxite/Alumina/Aluminium	100.0	100.0	100.0	100.0
Brazil	-	0.1	3.7	6.1
Guyana	18.0	12.3	9.7	6.3
Jamaica	58.1	69.0	42.1	36.0
Dominican Republic	3.9	2.4	1.1	0.3
Suriname	18.0	14.3	24.2	21.8
Venezuela	1.9	1.9	19.2	29.5
Copper	100.0	100.0	100.0	100.0
Bolivia	1.0	0.7	0.1	0.1
Chile	76.6	81.7	73.7	71.8
Mexico	0.6	0.9	5.2	9.1
Peru	21.8	16.7	20.9	19.0
Tin	100.0	100.0	100.0	100.0
Bolivia	95.9	90.9	87.5	79.5
Brazil	3.8	8.5	10.8	16.0
Peru	0.3	0.6	1.8	4.5
Iron ore	100.0	100.0	100.0	100.0
Brazil	40.9	55.6	76.6	81.2
Chile	13.5	12.2	6.5	6.3
Peru	12.4	5.6	3.3	4.2
Venezuela	33.2	26.6	12.9	7.5
Nickel	100.0	100.0	100.0	100.0
Cuba	100.0	58.1	73.8	91.5
Dominican Republic	-	41.7	26.2	8.4
Silver	100.0	100.0	100.0	100.0
Bolivia	14.2	11.5	37.4	11.4
Chile	7.2	4.3	38.0	25.1
Mexico	39.5	48.2	...	...
Peru	38.9	35.3	23.6	63.4
Lead	100.0	100.0	100.0	100.0
Bolivia	7.9	5.7	3.5	3.1
Mexico	27.4	35.4	15.4	10.6
Peru	64.1	58.8	80.9	86.3
Zinc	100.0	100.0	100.0	100.0
Bolivia	14.8	10.7	10.9	10.5
Mexico	36.3	38.7	32.0	15.8
Peru	48.9	49.8	57.0	73.2

Source: See table 16 of the Statistical Appendix.

Table 14

RELATIVE CHANGES IN LATIN AMERICA'S SHARE OF INTERNATIONAL TRADE

(Percentages with respect to the total value of world trade of each item)

Items/Years	Exports to: Latin America	Canada	European Economic Community	Centrally-planned economy countries	United States	Japan	Other developed countries	Other developing countries	Totals
I Exports Totals									
1970	0.96	0.32	1.60	0.32	1.92	0.32	0.32	-	5.77
1980	1.15	0.15	1.00	0.40	1.86	0.20	0.45	0.30	5.52
1983	1.20	0.16	0.98	0.60	1.80	0.27	0.44	0.44	5.88
II Exports of minerals and metals									
1970	0.45	0.13	2.65	0.36	2.15	0.81	0.46	0.10	7.11
1980	0.70	0.10	1.85	0.31	1.04	0.89	0.41	0.36	5.67
1983	0.48	0.13	1.69	0.42	1.28	1.13	0.34	0.58	6.05
III Exports of minerals									
1970	0.14	0.55	3.43	1.55	5.71	2.44	1.08	0.29	15.20
1980	0.35	0.27	3.90	1.01	2.77	2.64	1.44	1.15	13.54
1983	0.36	0.63	4.77	1.74	3.17	3.80	1.15	1.39	17.01
IV Exports of metals									
1970	0.54	-	2.42	0.01	1.11	0.33	0.27	0.05	4.74
1980	0.79	0.06	1.33	0.14	0.61	0.46	0.16	0.16	3.72
1983	0.51	0.02	1.00	0.12	0.86	0.53	0.16	0.40	3.59

Source: See tables 11, 12 and 13 of the Statistical Appendix.

greatest increases of minerals and metals exports from Latin America were achieved in those to Canada, the centrally-planned economy countries, the United States, Japan and other developed countries, whereas intra-regional exports and exports to the countries of the European Community and other developed countries declined. In the group of minerals exports alone, Latin America's share of world exports fell from 15.2% to 13.5% between 1970 and 1980 and increased to 17.0% in 1983. Between 1970 and 1983, this share increased in all cases at a lower rate save for exports to the United States. In the group of metals exports, the region's share declined, during the periods indicated, from 4.7% to 3.7% and 3.6%. The decline in this share, during the period 1970-1983, occurred in intra-regional exports and exports to the countries of the European Community, the United States and other developed countries whereas exports to Canada, the centrally-planned economy countries, Japan and other developing countries increased.

In terms of current prices, the share of minerals and metals exports of Latin America in total exports declined from the level of 18.4% that it had reached in 1978 to a mere 8.9% in 1982. This share fell in all the countries of the region except for Brazil, where it increased from 8.2% to 10.8%, Jamaica where it moved from 55.7% to 70.9% and Suriname where it showed an increase of 50.7% to 76.9% during the period 1970-1982 (see table 15). Except in the cases of Jamaica and Mexico this situation is creating relatively low ratios between the growth rates of mining production and the expansion rates of total exports.

The net extraregional exports of Latin America, in 1980, consisted of 12 products which accounted for 20% of extraregional exports for that year.

The share of minerals in those imports were as follows: potassium (41%), platinum (24%), phosphated rocks (13%), asbestos (12%), chromium (5%), magnesium (3%), manganese (1%) and other products with proportions lower than 1%: barite, fluorite, mercury, uranium and vanadium. The main importing countries were: Argentina, Brazil and Mexico. Following them in the number of imports were Chile, Colombia and Venezuela (see table 16).

For 1983, the following minerals exports have been estimated for Latin America: exports of minerals were estimated at around US$ 5.3 billion, almost 25% less than the figures for 1980 at constant 1983 prices and minerals imports (SITC 27, 28) were estimated at US$ 650 million with a decline 50% greater than the 1980 level and therefore generating a surplus of US$ 6.5 billion. Metals exports and imports (SITC, 67, 68, 13) were estimated at US$ 5 and 7 billion respectively, with a trade deficit of US$ 2 billion which is likely to be substituted, at least in part, by regional production. To this figure should be added the surplus created by imports of manufactured end use goods, produced from metals

Table 15

LATIN AMERICA: SHARE OF MINING EXPORTS a/ IN TOTAL EXPORTS b/

(In percentages)

Main Exporting Countries	1970	1974	1980	1982
Bolivia	77.3	57.9	58.5	43.9
Brazil	8.2	8.0	9.8	10.8
Chile	94.4	90.7	52.7	51.9
Guyana	53.7	33.3	43.8	34.2
Jamaica	65.7	67.1	76.4	70.9
Mexico	7.4	11.4	2.1	1.5
Peru	46.7	53.5	33.6	38.1
Dominican Republic	7.1	17.4	12.5	3.8
Suriname	50.7	38.7	82.2	76.9
Venezuela	7.1	2.7	3.4	3.9
Total 10 countries / Total Latin America	18.4	13.7	9.8	8.9

Source: See table 17 of the Statistical Appendix and ECLAC, Statistical Yearbook of Latin America, 1981 and 1983.

a/ Exports of minerals and metals of bauxite, copper, tin, iron ore, nickel, silver, lead and zinc.
b/ Exports of goods according to balance of payments' values, at current prices.

and minerals and which were estimated at 40% of total imports, with an amount higher than US$ 43 billion, over an amount of US$ 108 billion for total imports.

From the following figures, it can be observed that, except from iron ore and nickel, the ratio between the increase in world minerals exports and the increase in Latin American minerals production is very low. This situation could mean that regional production is reacting rather slowly to the changes in international trade and therefore causing significant changes in Latin America's share of world minerals and metals exports.

Table 16

LATIN AMERICA: ESTIMATE OF THE EXPORTABLE SURPLUS BY COUNTRIES - 1980
(In units of metric tons (MT))

Unit	Minerals	Argentina	Brazil	Chile	Colombia	Mexico	Peru	Venezuela	The other countries	Net exports	Latin America % of production
MT	Antimony	(690)	(2 155)		(30)	(121)	746	(30)	16 287	14 007	72.5
Thousands	Asbestos	(18)	(39)	(13)	(20)	(60)	(8)	(10)	(20)	(188)	(134.3)
Thousands	Barite	(4)	(8)	(1)		(1)		(54)	(8)	(76)	(6.5)
Thousands	Bauxite	(346)	2 388			(673)		(578)	20 765	21 556	85.6
MT	Bismuth	(23)		(26)	(3)	173	520		10	651	50.9
MT	Cadmium	19	(139)		168	319	9			376	39.5
Thousands	Chromium	(5)	(31)			(51)	(4)	(7)	30	(68)	(19.4)
MT	Cobalt	(138)	(281)		(5)	(54)		(10)	1 719	1 231	71.4
Thousands	Copper	(53)	(245)	1 028		49	348		(14)	1 113	69.1
Thousands	Tin		2			(2)	1		24	25	69.4
Thousands	Fluorite	(846)		(4)		1 380	(2)	(1)		(7)	(0.8)
Thousands	Iron ore		59 441	3 425			3 563	10 180		77 143	82.8
MT	Lithium			3 824						3 824	98.0
Thousands	Magnesium	(100)	(6)			(14)				(20)	
Thousands	Manganese	(60)	(14)		(3)	(38)		(2)	1	(158)	(5.6)
MT	Mercury	(2)	(187)		(30)	(15)		(6)	20	(278)	(380.8)
Thousands	Molybdenum		(2)	13		(1)	(2)			11	78.6
Thousands	Nickel	(9)	(9)						63	51	76.1
Thousands	Niobium		13			(3)				13	100.0
MT	Gold	(1)	29	(3)	8	6	1	(2)	15	53	67.1
MT	Silver	28	(120)	302	(117)	1 469	1 230	(15)	289	3 066	91.1
Thousands	Platinum	(8)			0.4		(41)	(1)		(49.6)	(12 400.0)
Thousands	Lead	(14)	(58)			48	125	(5)	(11)	90	24.3
Thousands	Potassium	(39)	(1 058)	(113)	(101)	(142)	(7)	(19)	(267)	(1 732)	(618.6)
Thousands	Phosphated rocks	(50)	(1 311)	(199)	(89)	(1 349)	(14)		(281)	(3 312)	(109.6)
Thousands	Rutile		0.4							0.4	100.0
MT	Selenium	(14)		233	(2)	6	76			289	90.9
MT	Tellurium	(1)		67		5	23			94	98.9
MT	Tungsten	(7)			(4)	(24)	283	(1)	3 359	3 606	66.3
MT	Uranium		(162)						(3)	(165)	(82.2)
MT	Vanadium		(557)	455	(24)					(126)	(27.7)
Thousands	Zinc	(69)				148	508	(26)	10	571	62.1

Source: See tables 5 and 10 of the Statistical Appendix.

	Annual growth rates of the volume of world mineral exports			Correlation coefficient
inerals	1961-1970 %	1970-1975 %	1975-1980 %	(R2)
auxite	6.23	3.44	2.61	0.34
opper	2.82	3.56	1.64	0.31
in	1.22	-1.11	-1.27	0.09
ron ore	8.88	3.38	-0.24	0.88
ickel	5.81	0.79	-0.99	0.84
ead	4.57	-1.09	0.67	0.25
inc	4.36	-2.50	0.15	0.04

As has been observed, the world crisis has affected the evels of production, consumption and international trade of he different mining products in different ways and in the ajority of cases this has obviously brought about changes in he trends of relative prices of the minerals. Comparing the istorical trends of the period 1947-1975 with those observed uring the period of the world crisis 1975-1982, the following lassification can be established for the different minerals see table 17):

a) <u>Trends that have not been affected during the crisis period</u>

i) Positive trends: arsenic, columbium, molybdenum, silver, tin and tellurium;

ii) Negative trends: ilmenite, vanadium, tungsten, asbestos, lead, antinomy and cadmium.

b) <u>Trends affected during the crisis period</u>

iii) From negative to positive: cobalt, barite, platinum, thorium, gold, lithium, sulphur, magnesium, bauxite, tantalum, manganese, mercury, copper and zinc.

iv) From positive to negative: potassium, fluorite, rutile, nickel, iron ore, phosphated rocks, chromium, selenium and bismuth.

Table 17

RELATIVE EVOLUTION OF THE INTERNATIONAL PRICES OF MINERALS a/

(Index: 1974 = 100)

1947		1965		1975		1980		1982		1983		First semester 1984	
Lithium	128	Uranium	233	Arsenic	329	Cobalt	667	Arsenic	571				
Thorium	100	Mercury	227	Uranium	233	Arsenic	457	Cobalt	460				
Arsenic	86	Thorium	150	Phosphated rocks	140	Silver	438	Columbium	389	Gold	266	Cobalt	304
Uranium	73	Asbestos	106	Manganese	127	Gold	383	Barite	250	Molybdenum	243	Gold	242
Ilmenite	59	Lithium	101	Fluorite	125	Molybdenum	379	Platinum	250	Silver	243		
Lead	56	Arsenic	86	Potassium	123	Columbium	333	Thorium	237				
Tungsten	56	Iron ore	84	Ilmenite	122	Thorium	288	Gold	235				
Manganese	55	Ilmenite	81	Iron ore	121	Tellurium	237	Molybdenum	228				
Fluorite	50	Rutile	76	Molybdenum	121	Platinum	231	Potassium	227				
Cobalt	46	Fluorite	75	Nickel	119	Potassium	207	Fluorite	225	Platinum	220	Platinum	211
Cadmium	44	Platinum	74	Rutile	119	Tin	205	Lithium	220	Potassium	207	Silver	197
Asbestos	41	Manganese	73	Lithium	118	Barite	200	Ilmenite	216	Magnesium	191	Nickel	184
Vanadium	40	Molybdenum	73	Sulphur	117	Fluorite	200	Rutile	203	Nickel	184		
Tantalum	36	Tellurium	72	Selenium	115	Nickel	197	Nickel	184				
Sulphur	33	Cadmium	63	Asbestos	114	Lithium	183	Sulphur	183				
Molybdenum	33	Copper	63	Thorium	112	Bauxite	178	Magnesium	179	Bauxite	170	Molybdenum	167
Platinum	32	Potassium	53	Vanadium	112	Ilmenite	178	Silver	178				
Mercury	31	Lead	52	Tellurium	111	Tantalum	172	Bauxite	169	Cobalt	161	Tin	158
Magnesium	27	Sulphur	50	Bauxite	109	Rutile	170	Tin	156	Tin	159		
Rutile	27	Columbium	50	Magnesium	109	Magnesium	167	Tantalum	156	Iron ore	153		

Table 17 (concl.)

1947	1965	1975	1980	1982	1983	First semester 1984
Copper 26	Bismuth 48	Columbium 106	Lead 152	Iron ore 142		Manganese 118
Barite 25	Tin 48	Tantalum 103	Sulphur 150	Mercury 141		Mercury 118
Zinc 23	Magnesium 47	Gold 101	Vanadium 149	Vanadium 134		Iron ore 105
Gold 22	Barite 45	Barite 100	Mercury 148	Tungsten 133	Antimony 137	Lead 104
Bismuth 21	Nickel 45	Chromium 100	Manganese 145	Tellurium 120	Manganese 127	Zinc 95
Tin 21	Cobalt 44	Tungsten 100	Iron ore 142	Asbestos 116	Mercury 117	Tungsten 89
Tellurium 21	Vanadium 42	Antimony 97	Asbestos 106	Phosphated rocks 100	Tungsten 100	Antimony 85
Nickel 20	Bauxite 33	Silver 94	Copper 106	Lead 93	Phosphated rocks 80	Phosphated rocks 80
Antimony 18	Tungsten 33	Bismuth 92	Phosphated rocks 100	Chromium 76	Copper 77	Copper 75
Columbium 17	Chromium 32	Platinum 86	Chromium 96	Copper 72	Lead 70	Selenium 63
Iron ore 16	Selenium 31	Tin 84	Antimony 84	Zinc 61	Zinc 62	Bismuth 47
Silver 15	Silver 27	Cadmium 82	Cadmium 82	Antimony 59	Selenium 24	Cadmium 40
Phosphated rocks 12	Antimony 25	Lead 70	Selenium 66	Cadmium 27	Cadmium 23	
Potassium 10	Zinc 25	Zinc 61	Zinc 63	Selenium 18	Bismuth 21	
Selenium 10	Gold 22	Copper 60	Bismuth 30	Bismuth 17		
Chromium 4	Phosphated rocks 20	Mercury 48				

Source: See table 19 of the Statistical Appendix.

a/ See in table 17 of the Statistical Appendix Latin America's major export and import minerals.

65

Chapter III

POTENTIAL FOR DEVELOPING MINING RESOURCES IN LATIN AMERICA

1. Evolution and medium-term prospects

As indicated in chapter I, there are two basic models by means of which the world economy could not only overcome the crisis period but also make the process of economic growth more dynamic. The first relates to the possibility of establishing a new international economic order, which would imply significant changes in the pattern of the international division of labour that existed in 1973. According to this pattern, the developing countries specialized in exporting their natural resources, whereas the developed countries' exports were mainly those products which required more capital and technology and human skills and, as was observed previously, this model is creating a crisis and limiting the opportunities for increased trade and therefore for greater economic development. Having regard to the high degree of international mobility of financial resources (capital) and of qualified human resources (know-how), the new pattern of the international division of labour would have to be based on the transfer of these resources to the developing countries, which would specialize in the exploitation and gradual industrialization of their natural resources, in the case of minerals, from the metallurgical and iron and steel producing phases to the processing of end use products with a high metal or mineral content. The developed countries, for their part, would specialize in high-technology industries with a low natural resource content. One of the main instruments that could initiate and gradually give shape to this new international division of labour would be the "improvement" of the long-term sales contracts which could include more equitable clauses for financing and the transfer of technology, including co-operation and technical advice and the provision of machinery and equipment and clauses providing for gradual industrialization which would imply changes in the features of the products to be marketed.

It is obvious that the change from one model to the other would have to be made gradually, using as a basis, on the one hand, the reactivation of the economy of the developed countries and on the other, the specific situation of each mineral, considering both its economic position and its medium-term trends.

International trade in minerals, metals and metallic products in terms of value is composed of the trade of the developed countries by as much as 70%. The United States trade accounts for 40% of that percentage, which explains the vital importance of the economy of that country. During the period 1963-1972, the annual growth rate of the gross domestic product (GDP) of the developed countries was, on the average, 4.7%. A similar rate was achieved during the period 1975-1980 (4.5%), but during the following period (1980-1982) this figure was lower than 1% and this level could only be compared with that of the years 1974 and 1975. The United States economy recovered quickly in 1983, the GDP rose to 6% and inflation was kept under control at a rate of around 4%. However, while the high interest rates made it possible, on the one hand, to issue public bonds in the amount of over US$ 200 billion, on the other hand, they curtailed the opportunities for long-term investments since the high growth rates of production were achieved basically by means of strong consumption expansion. The high interest rates also produced a significant revaluation of the dollar with the attendant increase in the trade balance deficit. The annual growth rate of United States production, which during the first quarter of this year was about 10%, fell to 7% during the second quarter and this downward trend continued during the third quarter, falling to less than 4%. During this quarter, industry continued operating at 83% of its capacity but housing construction fell by around 13%. As indicated above, the fundamental cause of this slowdown was the high cost of money, as a result of which, at the end of September, industries began to close down and it was hoped that by so doing, the growth of production will reach a rate of around 4.5% during the last quarter of this year. It is estimated that the production of the other developed countries will grow at the rate of 3% in 1984. In the case of Japan, this rate would be 4% compared with 3.5% in 1983, the Federal Republic of Germany would increase its production rate from 1.3% in 1983 to 2.5% in 1984 but in the United Kingdom it would probably decline from 3.5% to 2%.

With respect to Latin America's major potentially exportable products, according to the level of known reserves in 1981, the recent evolution and the medium-term outlook for these products could be the following:

- <u>Antimony</u>: The demand for antimony is closely connected to the evolution of the automobile industry and housing construction and demand was very low in 1983, but from January to May 1984, it increased at a steady rate, declining again in June. The market was controlled basically by production in South Africa, Bolivia and China. Its price trend is expected

Table 18

RELATIVE MEDIUM-TERM TRENDS IN MINERALS PRICES a/

(Minerals and levels) b/

Levels	Increasing trends	Levels	Constant trends	Levels	Declining trends
1	Columbium	2	Arsenic	9	Thorium
3	Barite	4	Cobalt	14	Potassium
6	Silver	5	Gold	15	Platinum
7	Lithium	11	Ilmenite	16	Molybdenum
8	Magnesium	12	Fluorite	19	Sulphur
10	Bauxite	13	Rutile	20	Tin
17	Tantalum	21	Nickel	27	Iron ore
18	Tellurium	24	Manganese	28	Lead
22	Zinc	25	Mercury	30	Tungsten
23	Vanadium	31	Copper	33	Asbestos
26	Selenium	34	Phosphated rocks		
29	Chromium	35	Antimony		
32	Bismuth	36	Cadmium		

Source: See table 19 of the Statistical Appendix.

a/ In keeping with the evolution of the numbers index with base 1974 for the years 1947, 1965, 1975, 1980, 1982, 1983 and the first semester of 1984.
b/ Place occupied by the index during the period 1982-1984.

to remain unchanged in the medium term (see table 18). Japan needs to import 100% of its consumption, the European Community 90%, the United States 51% and the USSR 20%. The main suppliers from the region would be Bolivia, Mexico and Peru (see table 8 of the Statistical Appendix).

- Bauxite-aluminium: Aluminium prices increased rapidly during the period 1978-1980 falling again in 1982. In 1983, there was a strong increase in the demand from the United States and Japan and the reductions in the production caused a drop in inventories and pushed up prices. However, the decline in prices during the first quarter of 1984 is creating a new situation of aluminium over-production. Depending on the performance of the inventories and on the basis of historic consumption trends, it could be estimated that aluminium and bauxite prices will tend to increase in the medium term. Despite this, while it is felt that the variations in the profitability of aluminium are dependent on energy and raw

69

material costs, it could also be assumed that there would be great pressure to keep bauxite prices low, and these would depend, in the final analysis, on the attitude adopted by the main producers: Jamaica, Australia, Guinea and Suriname. Japan depends on imports for 13% of its alumina, 31% of its aluminium and 100% of its bauxite. The import requirements of the EEC countries represent 84% of their total consumption of alumina and 28% of aluminium. The United States imports 94% of its bauxite requirements and the USSR 60%. The large reserves of the region could greatly increase the expansion of these exports, especially from Brazil, Guyana, Jamaica and Suriname.

- Copper: Whereas the demand for aluminium increased in 1983, copper remained depressed and was reactivated slightly because of the consumption of China and of a larger inventory formation in the refineries and this situation enabled prices to remain at levels similar to the 1982 levels. In 1984, prices recovered somewhat between January and April but this was wiped out by the declines in May and June. The decision of the President of the United States not to impose copper importation quotas, on the one hand, started a new period of low prices and on the other, created the possibility of achieving a stable balance between supply and demand. In the medium term, it is believed that, on the average, there would be some degree of price stability although it is predicted that an expansion of demand would make possible an increase in prices in the face of an inelastic supply and that this increase would be temporary since once again production would come from the marginal deposits, which would lead to a new level of over-production. The EEC depends on imports for 67% of its supplies, Japan 87% and the United States 5%. The region could cover a part of these requirements with exports from Chile, Mexico, Peru and possibly, from Panama.

- Columbium: The European Economic Community, the United States and Japan have to import all of their requirements of this mineral, which could partly be supplied by production from Brazil. It is believed that, in the medium term, prices would maintain an upward trend.

- Tin: Perhaps one of the major effects of the crisis was the reduction in tin consumption which, between 1978 and 1983, fell at annual rates of -3.3%. Although production also declined at similar rates in the five years indicated, production surpluses were followed by an increase in stocks. Notwithstanding this situation, the stabilizing action of the International Tin Council (ITC) caused prices to move upwards, from April 1984. This trend appears as if it will persist until the end of the year, as a result of an excess of consumption over production estimated at 15 000 metric tons. However, it must be borne in mind that the trade inventories now held by the producers and the ITC exceed 80 000 metric tons, to which should be added the strategic reserves of the General Services Administration (GSA) of the United States, estimated at over 167 000 tons. In light of this situation, it is believed that in the medium term, prices would tend to move upwards. Ninety-six per cent of Japan's requirements are met from imports, 95% of the EEC's, 80% of the United States and 11 of the USSR's. The region's main exporters are Bolivia and Brazil.

- Fluorite: In 1983, the downward trend in fluorite demand and prices continued but it is believed that recovery could begin at the end of this year and that it could be sustained without significant changes in the medium term, since the United States depends on imports for 85% of its supplies, the EEC 18%, Japan 100% and the USSR 47%. The main exporter in the region is Mexico but Argentina and Brazil also have some reserves.

- Iron ore and steel: During the 1970s and at the beginning of this decade, the price of iron ore was subject to rather wide fluctuations. In 1983, demand fell by 49% while prices declined by more than 11% and supply recovered considerably. It is estimated that, during the coming years, prices will maintain a declining trend and recover again during the period 1986-1987. After this period, it is expected that they will remain constant until the middle of the 1990s. Japan imports 99% of its total needs, the EEC 79% and the United States 28%. Latin America has the potential to extend its exports with production from Bolivia, Brazil and Cuba.

- Lithium: While demand was maintained without major variations, prices increased by 5% in 1983. On the other hand, whereas exports from China increased, exports from the USSR declined. New uses for lithium, especially in the specialized areas of electronics, medicine and photography lead one to predict a growing trend for prices in the medium term. Latin America has the potential to increase its exports of lithium with production coming particularly from Bolivia and Chile.

- Magnesium: Whereas the production of metallic magnesium grew by 8% in 1983, demand increased by 10%, the difference being covered by secondary production and the reduction of stocks by producers. In the medium term, it is expected that the upward trend of prices will continue. It is believed that Brazil's metal-producing plants are operating at slightly more than 20% of their productive capacity and that, therefore, production and exports could increase rapidly.

- Molybdenum: Consumption in 1983 was 35% lower than in 1979 and prices therefore fell considerably during that period. With the decline in production, the price recovered briefly in 1983, but the existence of large inventories pushed them down again in mid-1984 so that low prices are expected in both the short and medium term. The European Economic Community and Japan need to import 100% and 99% respectively to meet their needs.

- Nickel: For three years the demand for nickel was depressed but it increased by 10% in 1983 and a similar increase is expected this year. The gap between demand and production was filled by a reduction of inventories which at any rate enabled prices to recover. With the inevitable variations both positive and negative, the prices of nickel are expected at least to maintain a stable trend over the medium term. The EEC's import requirements of 80% and the United States requirements of 72% could be met partially by the large reserves in Brazil, Colombia, Cuba and the Dominican Republic.

- Phosphated rocks: Prices in the last few years increased until 1981. Subsequently, both demand and production stagnated until 1983, when demand increased by 12%, production by 9% and international trade by 6%. Despite the reactivation of demand, it would be difficult to reach the levels of installed capacity in the production of this mineral in the next few years and therefore, it is believed that prices would tend to remain stable in the medium term. Import requirements of 100% in the case of Japan and 99% in the EEC could in part be met by regional reserves, located mainly in Brazil, Mexico and Peru.

- Selenium: After several years of over-production and low prices, in 1983 demand grew by an estimated 29% and enabled prices to recover and from all appearances they will maintain a growing trend both in the short and medium term. All of the EEC's requirements and 49% of the United States could be supplied, to a large extent, by the large reserves in Chile, Mexico and Peru.

- Tantalum: With a market depressed since 1980, the 13% increase in demand, in 1983, reduced inventories and improved the prices of tantalum. It is expected that, in the medium term, demand will grow at high rates whereas the production of the main exporters, Malaysia and Thailand, would be restricted by production quotas of tantalum's co-product, tin. This situation leads one to believe that tantalum prices will move upwards in the medium term. The large import requirements of the United States (91%), of the EEC (100%) and of Japan (100%) could be partially met by production from Argentina and especially from Brazil.

- Tellurium: Both demand and prices depressed the tellurium market by 10% in 1983. However, the manifold uses of this product make one hazard that there will be a growing trend in the medium term. Exports from the region would come primarily from production in Peru.

- Titanium (ilmenite and rutile): Whereas in 1983, demand grew at around 6%, it is expected that in the medium term it will grow at annual rates of 5%. The stabilization of prices in the first half of 1984 leads one to expect that they would increase, or at least be stabilized both in the short and medium term. Reserves in Brazil could partially meet the import requirements of the European Economic Community (100%), Japan (100%) and the United States (43%).

- Vanadium: Consumption in 1983 fell to levels comparable with 1963 and this situation was made worse by China's export surplus. However, this situation changed in the last months of

1983 when a price recovery began. As the aeronautical industry is using significant amounts of vanadium, it is expected that in the medium term both demand and prices will show marked recovery. The import requirements of Japan and the EEC in the amount of 100% and of 42% in the United States could be partially met by supplies from Chile and Venezuela.

However, the action by the consumer countries to reduce dependency on imports of mining raw materials, the eagerness of the countries producing these raw materials to increase their added value and to process them industrially as one of their basic development options and the possible demise of the existing model of specialization and international trade would to a certain degree make the formation of highly self-sufficient mining economies more likely among large groups of countries: developed countries, centrally-planned economy countries and developing countries. It is obvious that these forms of autonomous growth, at the regional level, would require the appropriate specialization and national complementation since their dynamism would be based on high growth rates in international trade. On the other hand, the limits of self-sufficiency would be determined, in each region, by the mineral reserves available on the one hand, and on the other, by the expectations of consumption growth. The excess of consumption over production would determine the import requirements and outside the region, which would have to be met by an export surplus generated in another region. Using production, consumption and reserves known in 1981 as a base, projections have been made up to the year 2000, in order to determine the main features of a scenario as indicated above.

2. Projected minerals consumption in the year 2000

Latin America's share of world minerals consumption during the period 1980-1981 varied greatly from one mineral to the other and ranged from 0.2% for tellurium to 23.6% for platinum. In terms of consumption per capita the range ran from 2% to 84% of the consumption per capita of the developed countries in 1980 for uranium and manganese respectively. It should be borne in mind that this consumption refers to the industrial use of the mineral and not to its consumption or end use.

The following assumptions have been made for the estimate of consumption per capita and of total consumption in the year 2000.

a) Considering that in the developed countries, total consumption has reached a certain level of saturation, it is assumed that it would grow at annual rates of between 0.5% and 1.0% and that there would be few cases where this rate would fall below 0.5%. For the total consumption, and on the basis of the population growth between 1975 and 1980, it is assumed that there will be an annual population growth rate of 0.79% between 1980 and 2000.

b) The assumptions of the growth of consumption per capita of Latin America, other developing countries and centrally-planned economy countries are as follows:

i) That consumption per capita by the year 2000 would increase to 50% of the consumption levels reached by the developed countries in 1980, for those products which represented up to 15% of that year's consumption.

ii) That those products which, in 1980, represented between 16% and 50% of the levels of consumption per capita of the developed countries would increase to 75% of the 1980 levels.

iii) That similarly, the products which, in 1980, represented 51% of the consumption per capita of the developed countries, would increase to 100% of the 1980 levels.

iv) That the products which, in 1980, represented more than 75% of the consumption per capita of the developed countries would increase up to 100% of those levels in the year 2000. Finally, the appropriate adjustments would be made, when the total consumption represented more than 100% of the amount of the estimated world reserves in 1981.

c) In estimating the total consumption of these three groups of countries, it is believed that based on the growth during the period 1975-1980, annual population growth rates would be as follows: Latin America 2.43%, other developing countries 1.69% and the centrally-planned economy countries 1.31%.

In accordance with these assumptions, consumption levels in the year 2000 have been estimated for each group of countries (see table 20 of the Statistical Appendix) and world consumption is expected to grow at annual rates ranging from 2% for tin to 7.8% for vanadium and bauxite. The growth of non-ferrous traditional metallic minerals is expected to fluctuate between 2% for tin and 3% for zinc. The range in the traditional ferrous metallic minerals would be from 2.9% for tungsten to 7.8% for vanadium and in the other light metals the range would be from 5.7% for titanium (ilmenite) to 7.8% for bauxite (see table 19).

While it appears that the estimated growth rates are very high, especially in Latin America, it should be borne in mind that these rates are relatively low for the more important traditional minerals and that possibly the consumption levels of the base year would be much higher if the calculation included minerals used in the imports manufactured with metal and mineral inputs.

Table 19

ESTIMATED PER CAPITA CONSUMPTION OF MINERALS BY THE YEAR 2000

(In kilograms)

Minerals	Projected rate %	Category a/	High annual growth rates 1980-2000 Estimated per capita consumption			
			Latin America	Other developing countries	Developed countries	Centrally-planned economy countries
Vanadium	7.85	II	0.02	0.02	0.03	0.03
Bauxite	7.79	VI	82.30	70.23	90.93	61.73
Tellurium	7.06	IV	0.28	0.20	0.61	0.41
Lithium	6.48	V	3.10	2.60	6.84	4.64
Rutile	6.16	VI	0.26	0.24	0.28	0.26
Selenium	6.16	IV	1.02	0.30	2.25	1.53
Phosphated rocks	6.14	V	57.04	56.01	126.04	85.55
Molybdenum	6.06	II	0.06	0.01	0.13	0.09
Magnesium	5.99	VI	0.12	0.07	0.25	0.23
Platinum	5.77	VII	0.15	0.14	0.15	0.15
Ilmenite	5.73	VI	1.36	1.33	2.20	1.49
Cadmium	5.70	IV	7.80	3.63	17.23	11.69
Fluorite	5.23	III	2.07	2.04	2.60	2.33
Cobalt (g)	5.18	II	14.00	6.66	33.96	18.10
Nickel	5.06	II	0.32	0.08	0.70	0.47
Chromium	5.05	II	4.01	3.92	6.43	4.37
Antimony	4.68	V	31.91	16.07	47.01	31.91

MINERALS	Projected rate %	Category	Lower annual growth rates 1980-2000 Estimated per capita consumption			
			Latin America	Other developing countries	Developed countries	Centrally-planned economy countries
Iron ore	4.21	II	172.52	89.86	381.22	258.77
Potassium	4.14	V	9.66	4.99	21.35	14.49
Barite	4.03	II	3.78	2.84	4.18	2.18
Uranium	4.03	VIII	0.03	0.01	0.06	...
Manganese	3.98	II	11.09	8.45	11.09	11.09
Mercury	3.63	IV	1.73	1.01	3.81	3.44
Gold	3.58	VII	0.37	0.16	0.82	0.56
Bismuth	3.45	V	1.27	0.67	1.88	1.40
Zinc	3.31	I	1.61	0.99	5.12	2.00
Tungsten	2.88	II	14.75	7.10	31.99	20.49
Copper	2.62	I	2.00	1.41	7.95	3.00
Lead	2.39	I	1.07	0.79	4.24	1.18
Silver	2.15	VII	2.27	1.46	5.01	3.41
Asbestos	2.09	III	1.10	0.32	2.55	1.91
Tin	2.01	I	0.05	0.02	0.14	0.07

Source: See table 20 of the Statistical Appendix.

a/ Categories:
I   Traditional non-ferrous metals   V   Chemical use minerals and metals
II Traditional ferrous metals   VI   Light metals
III Insulators and refinements   VII Precious metals
IV Electrical use metals   VIII Radioactive metals

3. Prospects for minerals production in the year 2000

Estimates of production in the year 2000 for the groups of countries have been made on the basis of the following assumptions:

a) That the exhaustion in the year 2000 of known reserves in 1981 has been taken as a maximum limit of production expansion (see table 6 of the Statistical Appendix).

b) That production of the other minerals in the year 2000 would be equal to the consumption levels for each group of countries.

c) That in the first case, the consumption deficit (extraregional imports) would be covered by that group of countries which had the better reserves to production ratio.

d) According to the United Nations statistical classifications, the first group includes the countries of Latin America and the Caribbean; the second group includes all the developed countries not included in the other groups; the group of developed countries includes the Western European countries (excluding Yugoslavia), Oceania, Canada, United States, Japan and South Africa; the group of centrally-planned economy countries would include the Eastern European countries, including Yugoslavia, the People's Republic of China and the Democratic Republic of Korea.

The estimate based on the following assumptions would yield the following results in the year 2000 (see table 20):

a) At the world level, growth rates of production would be equal to estimates for the consumption of each product. In the year 2000, the reserves known in 1981 of the following products would be exhausted: asbestos, barite, bismuth, cadmium, mercury, gold, silver and lead.

b) In Latin America, production would grow at annual rates ranging from -4% for gold and 25.2% for uranium. The products whose reserves,· according to production forecasts, would be exhausted by the year 2000 are: asbestos, barite, bismuth, cadmium, cobalt, chromium, ilmenite, manganese, mercury, gold, silver, platinum, lead, potassium, rutile, tungsten and zinc.

c) In the group of other developing countries, the production rates would fluctuate between -4.3% for silver and 29.2% for manganese. By the year 2000, the reserves of the following products known in 1981 would be exhausted: antimony, asbestos, barite, bismuth, cadmium, fluorite, iron ore, manganese, mercury, gold, silver, lead, tungsten, uranium, vanadium and zinc.

d) In the group of developed countries, the production rates would vary between -0.2% for manganese and 10.2% for cobalt. By the year 2000, the reserves of the following products would have reached critical levels: asbestos, barite, bismuth, cadmium, cobalt, mercury, gold, silver and lead.

e) The group of centrally-planned economy countries would have annual rates varying from 0.0% for asbestos to 19.6% for

Table 20

ESTIMATED MINERALS PRODUCTION IN THE YEAR 2000
(Annual growth rates 1980-2000)

Minerals	Estimated production rates (%)				Year of depletion of the reserves known in 1981				
	Latin America	Other developing countries	Developed countries	Centrally-planned economy countries	World	Latin America	Other developing countries	Developed countries	Centrally-planned economy countries
Antimony	2.48	6.13	4.00	6.29	2 018	2 009	2 000	2 006	2 038
Asbestos	5.68	2.28	3.57	-0.01	2 000	2 000	2 000	2 000	2 001
Barite	-0.34	1.71	8.68	3.45	2 000	2 000	2 000	2 000	2 000
Bauxite	3.37	11.41	7.11	8.04	2 071	2 160	2 089	2 047	2 000
Bismuth	1.37	7.43	4.77	1.15	2 000	1 998	1 998	1 998	1 998
Cadmium	5.65	11.61	4.77	0.59	2 000	2 001	1 999	2 000	2 000
Cobalt	16.82	-0.65	10.17	11.79	2 040	2 002	2 112	2 000	2 027
Copper	1.07	0.94	5.11	4.66	2 025	2 113	2 060	2 007	2 000
Chromium	1.96	7.74	2.35	5.38	2 177	2 000	2 128	2 590	2 010
Tin	2.35	-1.48	3.68	6.67	2 015	2 015	2 034	2 005	2 003
Fluorite	4.76	10.66	5.08	1.36	2 014	2 012	2 000	2 030	2 000
Iron ore	7.15	-3.56	2.97	5.38	2 122	2 285	2 000	2 096	2 070
Ilmenite	-	13.98	0.43	13.34	2 046	2 000	2 011	2 129	2 010
Lithium	20.48	-	1.12	...	2 127	2 165	2 140	2 043	...
Magnesium	-	29.28	-0.24	8.81		-	-	6 463	...
Manganese	3.15	3.15	5.78	2.95	2 023	2 000		2 138	2 020
Mercury	11.47	1.68	4.45	2.67	2 000	1 998	2 000	1 998	2 000
Molybdenum	10.60	21.82	3.29	8.49	2 021	2 033	1 997	2 025	1 999
Nickel	5.71	0.96	3.39	7.72	2 032	2 171	2 012	2 021	1 999
Gold	-4.57	7.90	3.54	3.20	2 000	1 996	2 243	1 997	2 019
Silver	0.58	-4.34	3.75	2.45	1 998	1 997	1 997	1 997	2 000
Platinum	8.33	-	5.06	8.98	2 044	1 995	1 997	2 096	1 999
Lead	2.53	2.23	5.71	2.97	2 002	2 002	-	2 003	2 000
Potassium	-	14.45	1.08	4.57	2 173	2 003		2 251	1 999
Phosphated rocks	12.61	5.99	3.46	8.43	2 223	2 050	2 028	2 137	2 203
Rutile	13.83	10.39	0.59	19.16		1 999	2 502		2 071
Selenium	6.70	7.26	3.99	9.19	2 033	2 055	2 050	2 029	
Tellurium	2.64	9.33	5.30	9.20	2 028	2 012	2 043	2 032	2 000
Tungsten	2.97	2.73	3.85	2.28	2 015	2 000	2 002	2 006	2 000
Uranium	25.29	4.23	2.70	...	2 017	2 012	2 000	2 025	2 029
Vanadium	18.25	4.89	4.55	9.65	2 149	2 011	2 004	2 263	2 132
Zinc	0.48	4.56	4.18	1.90	2 006	2 000	2 000	2 010	2 000

Source: See table 21 of the Statistical Appendix.

77

Table 21

ESTIMATED EXPORTABLE SURPLUS IN THE YEAR 2000
(In units of metric tons (MT))

Unit	Minerals	Totals Production=consumption	Latin America Production	Latin America Consumption	Latin America Exportable surplus	Other developing countries Production	Other developing countries Consumption	Other developing countries Exportable surplus	Sub-total Exportable surplus
Thousands	Antimony	161	31	18	13	23	39	(16)	(3)
Thousands	Asbestos	7 408	421	629	(208)	617	773	(156)	(364)
Thousands	Barite	16 696	1 097	2 154	(1 057)	4 725	5 273	(548)	(1 605)
Thousands	Bauxite	418 695	46 911	46 911		170 171	170 171		
MT	Bismuth	6 746	1 679	724	955	390	1 620	(1 230)	(275)
MT	Cadmium	51 435	4 935	4 446	489	9 000	8 795	205	694
Thousands	Cobalt	90	4	8	(4)	20	16	4	
Thousands	Copper	15 799	1 772	1 140	632	2 306	1 658	648	1 280
Thousands	Chromium	26 053	516	2 286	(1 770)	9 501	9 501		(1 770)
Thousands	Tin	331	56	27	29	109	50	59	88
Thousands	Fluorite	12 978	2 332	1 179	1 153	4 272	4 936	(664)	489
Thousands	Iron ore	1 161 259	281 998	98 336	183 662	34 068	217 730	(183 662)	
Thousands	Ilmenite	8 880	80	778	(698)	3 230	3 230		(698)
Thousands	Lithium	23	11	2	9	6	6		9
Thousands	Magnesium	909	68	68		170	170		
Thousands	Manganese	58 258	5 254	6 321	(1 067)	7 671	20 474	(12 803)	(13 870)
MT	Mercury	13 522	640	984	(344)	1 367	2 438	(1 071)	(1 415)
Thousands	Molybdenum	350	108	34	74	24	24		74
Thousands	Nickel	1 914	182	182		190	190		
MT	Gold	2 442	31	213	(182)	325	396	(71)	(253)
MT	Silver	15 948	3 775	1 294	2 481	366	3 533	(3 167)	(686)
MT	Platinum	866	4	88	(84)		339	(339)	(423)
Thousands	Lead	8 647	610	610		490	1 914	(1 424)	(1 424)
Thousands	Potassium	64 927	5 506	5 506		12 091	12 091		
Thousands	Phosphated rocks	447 647	32 513	32 513		135 705	135 705		
Thousands	Rutile	1 477	6	146	(140)	575	575		(140)
MT	Selenium	6 302	1 164	581	583	1 622	727	895	1 478
MT	Tellurium	1 989	160	160		840	485	355	355
MT	Tungsten	94 163	7 820	8 407	(587)	13 881	17 195	(3 314)	(3 901)
Thousands	Uranium	97	17	17		21	24	(3)	(3)
Thousands	Vanadium	154	13	13		6	56	(50)	(50)
Thousands	Zinc	11 823	1 011	920	91	798	2 399	(1 601)	(1 510)

Table 21 (concl.)

Unit	Minerals	Developed countries			Centrally-planned economy countries			Sub-total
		Production	Consumption	Exportable surplus	Production	Consumption	Exportable surplus	Exportable surplus
Thousands	Antimony	46	43	3	61	61		3
Thousands	Asbestos	3 973	2 315	1 658	2 397	3 691	(1 294)	364
Thousands	Barite	7 792	3 791	4 001	3 082	5 478	(2 396)	1 605
Thousands	Bauxite	146 436	82 474	63 962	55 177	119 139	(63 962)	-
MT	Bismuth	4 158	1 703	2 455	519	2 699	(2 180)	275
MT	Cadmium	33 000	15 628	17 372	4 500	22 566	(18 066)	(694)
Thousands	Cobalt	31	31		35	35		
Thousands	Copper	7 211	7 211		4 510	5 790	(1 280)	(1 280)
Thousands	Chromium	5 837	5 832	5	10 199	8 434	1 765	1 770
Thousands	Tin	35	123	(88)	131	131		(88)
Thousands	Fluorite	4 243	2 358	1 885	2 131	4 505	(2 374)	(489)
Thousands	Iron ore	345 767	345 767		499 426	499 426		
Thousands	Ilmenite	2 693	1 995	698	2 877	2 877		698
Thousands	Lithium	6	6			9	(9)	(9)
Thousands	Magnesium	227	227		444	444		
Thousands	Manganese	23 929	10 059	13 870	21 404	21 404		13 870
MT	Mercury	7 160	3 453	3 707	4 355	6 647	(2 292)	1 415
MT	Molybdenum	154	118	36	64	174	(110)	(74)
MT	Nickel	635	635		907	907		
MT	Gold	1 572	748	824	514	1 085	(571)	253
MT	Silver	8 148	4 548	3 600	3 659	6 573	(2 914)	686
MT	Platinum	298	140	158	564	299	265	423
Thousands	Lead	5 649	3 846	1 803	1 898	2 277	(379)	1 424
Thousands	Potassium	19 364	19 364		27 966	27 966		
Thousands	Phosphated rocks	114 318	114 318		165 111	165 111		
Thousands	Rutile	396	256	140				140
MT	Selenium	2 041	2 041		1 475	2 953	(1 478)	(1 478)
MT	Tellurium	553	553		436	791	(355)	(355)
Thousands	Tungsten	31 629	29 015	2 614	40 833	39 546	1 287	3 901
Thousands	Uranium	59	56	3	...	...		3
Thousands	Vanadium	48	23	25	87	62	25	50
Thousands	Zinc	7 617	4 644	2 973	2 397	3 860	(1 463)	1 510

Source: See tables 20 and 21 of the Statistical Appendix.

rutile. The products with critical levels of reserves by the year 2000 would be asbestos, barite, bauxite, bismuth, cadmium, copper, tin, fluorite, mercury, molybdenum, gold, silver, platinum, lead, selenium, tellurium and zinc.

In this scenario of mining production, a number of cases may have been presented which seem to have a far from logical development as in the case of production and high reserve inventories (e.g., copper in Latin America) or high growth rates of production with rapid exhaustion of known reserves in 1981 (e.g., rutile in Latin America), but these cases are submitted in order to respect the methodological assumption that each group of countries would achieve the highest degree of self-sufficiency.

Despite the high rates estimated for the growth of some products, the total mining production of Latin America at constant 1975 prices would grow at an annual rate of 3.6% between 1980 and 2000 (see table 22 of the Statistical Appendix). This rate is lower than that for the historical period of the last decades and might be less than that of the total output of the region during the period 1980-2000, and therefore the share of mining output in the extractive phase would be declining. On the other hand, it could be expected that the output during the following stages of industrialization of minerals would increase.

The industrialization of the mining resources of the region would have the basic prerequisite for an integrated industrial structure in respect of inputs and would be complementary in respect of end use products. It is obvious that this process of industrialization would require financial resources and technology which do not exist in the region and that therefore appropriate co-operation would be needed from the international community, perhaps in associated forms (joint-ventures) of vertically integrated production and intra-regional marketing.

4. Estimates of exportable surpluses by the year 2000

As indicated previously, the known reserves in 1981 would hamper progress towards self-sufficiency in certain minerals, creating a deficit or a considerable need at the regional level which would be covered by the exportable surpluses from another group of countries (see table 21). In the case of Latin America, it could import these minerals from the following sources:

- Cobalt, chromium, ilmenite, rutile from the other developing countries.
- Asbestos, barite, manganese, mercury, gold, platinum and tungsten from the group of developed countries.

It should be borne in mind that given its mining potential, the region could, through an adequate exploration programme, succeed in replacing these import supplies by regional production.

80

In turn, Latin America would generate the following exportable surpluses for sale to the other groups of countries:

- Antimony, bismuth, fluorite, iron ore, molybdenum, silver and zinc to be sold primarily to the group of other developed countries.
- Tin, to be sold to the group of developed countries.
- Cadmium, copper, fluorite, lithium and selenium to be sold to the group of centrally-planned economy countries.

According to the above estimates, the following qualitative changes would take place in the structure of Latin America's extraregional trade in minerals by the year 2000, compared with the situation in the year 1980 (see table 22):

a) The exportable surpluses in the cases of bauxite, nickel, lead, tellurium and tungsten would be depleted.

b) Whereas in 1980 there were exportable surpluses of cobalt, gold and rutile, in the year 2000, these metals would have to be imported.

c) Imports made in 1980 would be replaced, in the cases of fluorite, magnesium, potassium and phosphated rocks.

d) New import requirements of ilmenite would be created.

In quantitative terms, the net exports from the region in 1980 represented 80.2% of the total mineral exports, but by the year 2000, this surplus would be reduced to only 30.4% in terms of value at constant 1975 prices. Whereas the trade surplus with other developing countries would increase, there would be balanced trade with the group of centrally-planned economy countries and a large deficit in minerals trade with the group of developed countries.

On the assumption that, in the long term, minerals prices maintain a close correlation with the evolution of production costs and levels, an estimate has been made on the price index by the year 2000, in terms of the previously estimated production levels and of the elasticity coefficients calculated for the period 1947-1974. According to this estimate, the minerals whose prices would have a more favourable evolution in the year 2000 would be cobalt, platinum, rutile, tellurium and uranium. On the other hand, there would be less favourable indices of evolution for antimony, asbestos, bismuth, cadmium and nickel. However, in terms of the possible evolution of the indices of production and prices, the index of total income (value of production) would evolve very favourably in the cases of bauxite, cobalt, magnesium, platinum, rutile, tellurium and vanadium (see table 23).

Using the methodology indicated, an estimate has been made of the exportable surplus and the minerals import requirements in the context of the intra-regional trade of Latin America in the year 2000. This trade, which in 1980 represented 2.6% of the region's mining exports (see table 13 of the Statistical Appendix), in the year 2000 would reach a

Table 22

LATIN AMERICA: ESTIMATED CHANGES IN THE COMPOSITION OF EXTRA-REGIONAL MINERALS TRADE
(Millions of 1975 dollars)

Minerals	1980		2000		Countries of destination (or origin) in 2000		
	Exports	Imports	Exports	Imports	Other developing countries	Developed countries	Centrally-planned economy countries
Antimony	57		52	165	52	(165)	
Asbestos		148		47		(47)	
Barite		3					4
Bauxite	495		18		18		
Bismuth	12	4					
Cadmium	4			33	(33)		
Cobalt	10						
Copper	2 281	69	1 296	1 950		(6)	1 296 / (1 994)
Chromium	205		238			238	
Tin			101		58		43
Fluorite	1 456	1					
Iron ore			3 490	492	3 490	(492)	
Ilmenite	79		186				186
Lithium							
Magnesium		33		117		(117)	
Manganese	59	17		3		(3)	
Mercury	196	2	397				397
Molybdenum	272						
Nickel	465			935		(935)	
Gold			376		376		
Silver	53	302		514		(514)	
Platinum							
Lead	1						
Potassium		520					
Phosphated rocks		166		114		(140)	
Rutile	12		3				3
Selenium	2						
Tellurium	1						
Tungsten			113		113		
Zinc	705						
Total	6 365	1 261	6 274	4 370	4 107/(33)	238/(2 419)	1 929/(1 994)
Surplus or net deficit as a percentage of exports	80.2		30.4		99.2	(916)	(3.4)

Table 23

ESTIMATED PRICE INDEX OF MINERALS IN THE YEAR 2000

(Base year 1974=100)

Minerals	Production Index	Elasticity 1947-1974	Price Index	Price index World Bank estimates	Index of value of production
Antimony	224	0.80	179	329	401
Asbestos	161	0.51	82	-	132
Barite	372	1.30	484	-	1 800
Bauxite	545	0.76	414	450	2 256
Bismuth	140	0.79	111	484	155
Cadmium	270	0.40	108	-	292
Cobalt	277	3.33	922	-	2 555
Copper	196	1.43	280	357	549
Chromium	351	0.68	239	-	838
Tin	141	1.64	231	570	326
Fluorite	268	1.37	367	-	984
Iron ore	129	3.12	402	421	519
Ilmenite	316	1.20	379	-	1 198
Lithium	329	1.01	332	-	1 093
Magnesium	689	0.71	489	-	3 370
Manganese	256	1.02	261	387	668
Mercury	147	1.82	268	-	393
Molybdenum	407	1.12	456	570	1 855
Nickel	242	0.86	208	220	504
Gold	177	1.45	257	-	454
Silver	157	1.79	281	-	441
Platinum	435	1.54	670	-	2 914
Lead	224	1.64	367	484	823
Potassium	246	1.72	423	-	1 041
Phosphated rocks	366	1.08	395	-	1 447
Rutile	446	1.45	647	-	2 884
Selenium	513	0.66	339	-	1 738
Tellurium	580	3.00	880	-	5 104
Tungsten	254	1.23	312	-	794
Uranium	285	2.00	570	-	1 625
Vanadium	811	0.53	430	-	3 486
Zinc	188	1.30	244	329	459

Source: 1. See tables 12, 19 and 21 of the Statistical Appendix.

2. World Bank office memorandum, Half yearly revisions of commodity price forecasts and quarterly review of commodity markets, 20 January 1984.

proportion close to 15%. The countries with the greatest intra-regional import requirements would be _Argentina_ for almost all products except cadmium, tin, fluorite, manganese, nickel and tungsten; _Mexico_ for almost all products except for mercury, silver, lead and tellurium; _Venezuela_ with import requirements of antimony, barite, bauxite, cobalt, chromium, fluorite, iron ore, ilmenite, manganese, nickel, gold, silver, potassium, phosphated rocks, rutile and zinc. The countries with the largest exportable surpluses would be: Bolivia, Chile, Colombia, Cuba, Jamaica and the Dominican Republic. There would be a situation of virtual equilibrium in Brazil and Peru, whilst other countries would face a deficit situation in the year 2000 (see tables 23 and 24 of the Statistical Appendix).

In view of the requirements for minerals substitution, it is estimated that production expansion in extraction and concentration of minerals would have an investment requirement of about US$ 65 billion at 1975 prices. If it is considered that this investment should be concentrated over the next ten years, because of the region's external debt situation, it is felt that the necessary resources coming from outside should, in turn, be concentrated over the first years of the decade indicated. For the whole period, it is estimated that these resources should represent at least 20% of total investments. This calculation does not include investments in the production phases of metallurgy, subsequent industrial processing of these products and the investments for marketing.

Finally, it must be remembered that even with the scenario of self-sufficiency, the exclusively mining sector would generate, annually, a net foreign exchange surplus for the region equivalent to US$ 3 billion at 1975 prices (see table 24).

Table 24

LATIN AMERICA: ESTIMATED MINING ACTIVITY IN 2000 a/

(Millions of 1975 dollars)

Countries	Industrial consumption	Production	Import substitution	Intra-regional trade Imports	Intra-regional trade Exports	Extra-regional exports	Investment requirements	% of internal financing	Net flow of annual foreign exchange 2000	Share of the GDP in the year 2000 %	Share of exports in the year 2000 %
Argentina	1 176	286	49	879	32	6	850	90	(884)	0.4	0.3
Bolivia	313	1 671	-	116	710	764	3 400	60	1 188	23.2	100.0
Brazil	7 583	7 195	2 776	633	614	2 407	30 000	90	888	2.8	7.0
Chile	954	4 228	-	372	2 385	1 261	8 500	60	2 849	21.0	43.3
Colombia	489	127	526	-	131	33	3 200	60	4	1.9	4.6
Cuba	491	872	-	187	305	263	1 800	60	291	...	...
Jamaica	101	448	-	38	385	-	900	60	302	...	...
Mexico	3 527	1 801	976	1 105	43	312	9 000	90	(1 200)	1.1	1.6
Peru	1 163	2 200	19	731	814	973	4 500	75	831	5.9	35.4
Dominican Republic	58	182	-	11	132	3	600	60	94	1.9	6.1
Venezuela	732	439	-	543	15	235	350	60	(310)	0.6	2.1
Other countries	1 420	462	24	1 164	213	17	1 100	60	(989)	0.8	1.7
Latin America	18 007	19 911	4 370	5 779	5 779	6 274	64 200	80	3 064	2.6	9.7

Source: See tables 22, 23 and 25 of the Statistical Appendix.

ECLAC: Economic Projections Centre: Macroeconomic model: The development of Latin America, its evaluation and long-term prospects.

Preliminary figures at 5 October 1984, on the basis of the following assumptions: the GDP in 1990 will maintain the per capita production
level of 1980 and the GDP in 2000 will have a growth rate of 3.5% of the per capita production between 1990 and 2000.

a/ Excluding Metallurgy and Iron and Steel Works.

Chapter IV

CONCLUSIONS: NOTES ON A NEW POLICY FOR DEVELOPING THE MINING RESOURCES OF LATIN AMERICA

1. The short to medium term

During the period 1960-1980, the annual growth rates of consumption or industrial use of minerals in Latin America were double the production rates, reducing the proportion of exportable surplus and generating for a group of minerals, extraregional import requirements which in 1980 represented 20% of the value of the regional exports of these products. In other cases, the margin of the exportable surplus represented between 15 to 85% of regional production.

One of the main factors limiting production in one group of minerals has been the low levels of investment in prospecting and mining exploration. If the consumption, production and investment trends are maintained, the region would have virtually exhausted its known reserves in 1981 by the year 2000, in a group of 17 important minerals (asbestos, barite, bismuth, cadmium, cobalt, chromium, ilmenite, manganese, mercury, gold, silver, platinum, lead, potassium, rutile, tungsten and zinc). However, the region has not only huge reserves of the other group of minerals but also potential reserves which are still not sufficiently known because of the lack of investment in exploration in the field, which would make it possible on the one hand, to achieve regional self-sufficiency and on the other hand, to generate export surpluses of significant size to be sold on the international market.

Another of the factors limiting mining production in Latin America, especially in the last few years, has been the rather undynamic evolution of foreign trade in minerals. During the period 1970-1974, exports from the region grew at an annual rate of 2.8% so that the region had a share of 15.2% of world minerals exports. During the period 1970-1984, that rate was negative (-0.7%) and the region's share fell to 13.5% and experienced a less favourable evolution than the rest of the world. During the last period 1980-1983, although it had a negative evolution with an annual rate of -2.7%, its share of world exports increased to 17%.

In 1983, extraregional exports of minerals reached US$ 5.3 billion, which at current prices represented a decline of 25% compared with 1980. The extraregional imports were US$ 650 million with a decline of 46% compared with the level in 1980. On the other hand, exports and imports of metallic products and semi-processed products had values of US$ 5 and US$ 7 million respectively, with a deficit of US$ 2 billion for the region. Furthermore, the imports of processed products with a metal or mineral base, especially capital goods, reached US$ 43 billion, representing around 40% of Latin America's total imports.

Latin America's mining exports are composed of a score of products, eight of which represent more than 90% of the value of those exports. The unfavourable evolution of the export volumes and the prices of minerals determined that their share at current prices fell from 18.4% to 8.9% of the total exports of the region between 1970 and 1983 and this situation was due largely to the low industrial profit index which Latin America's mining exports have.

The foregoing facts point to the need for a decision on a mining policy which should include the following:

a) improvement of the bargaining power of the region by creating an export structure which was more responsive to the changes in the international market;

b) a more detailed study of the possibilities for replacing the imports of minerals, metallic products and goods manufactured with a metal or mineral base;

c) the diversification of the production and mining exportation structures;

d) the co-ordinated management of supply inventories;

e) greater levels of profit and industrial processing of minerals.

2. The medium to long term

Even before the world economic crisis in the 1970s, another crisis was well under way, namely, the crisis of non-renewable natural resources and one of its main forms of expression was the oil pricing policy imposed by OPEC at the end of 1973. The truth is that this crisis began or was heightened at the end of the Second World War, because of the existence of a new international division of labour between the countries which exported manufactures and the countries which exported raw materials. Nevertheless, the mining crisis is not only characterized by the difference between the objectives and interests of the exporters and importers of minerals but also by the need to integrate mining activities with those of metallurgy and industries of end use goods, a necessary process for the mining output to reach the consumer. During the first half of the century, this process was made easier by the activities of transnational enterprises of production which had mining concessions in the raw materials exporting countries, as well as metallurgical and industrial plants in the countries that produced manufactured goods. Although these activities were not integrated into the national economy, the process was underpinned by the concentration of productive factors in the hands of the producing transnational enterprises. The small share of the mineral exporting countries in the profits of this industrial process, subsequently led to the strong desire not only to exercise full rights of sovereignty over their natural wealth but also to participate directly in that production process. The successive nationalizations of mining enterprises led to the segmentation of the productive process, whereby mining-metallurgical producers were located in the minerals exporting countries and the metallurgical producers and producers of manufactured goods with a mineral base were located in the importing countries. If it is further remembered that the prices on the international market were quoted for metals which are homogeneous products and not for metals which are differentiated products, it could be argued that marketing agents or intermediaries were needed to "integrate" the mining products at the level of metallurgical products or of the manufactured goods. The action of the marketing agents made the market system more complex since supply was not shaped by production alone but also by the liquidation of commercial inventories and demand was shaped by the consumption and the formation of those inventories. This speculative action by the marketing intermediaries pleases neither producers of minerals nor metalworkers nor industrial users of metals since this is one of the main factors creating short-term price distortions. Hence the need to return to integrating the productive process in mining and industry and for this, the following alternatives are suggested:

a) The producers-exporters of mining raw materials could form their own marketing enterprise. To this end, it would be useful during a first stage to join existing marketing enterprises where the necessary technical personnel could be trained.

b) The producers-exporters of minerals, through production could contribute to the work of the metallurgical and industrial productive enterprises currently situated in the countries which import mining raw materials.

c) In the minerals-exporting countries, high-efficiency integrated plants could be established with the participation (joint-ventures) of the metallurgical and industrial enterprises of the countries which currently need to import raw material. This alternative would imply a process of industrial redeployment, which would generate changes favourable to the developing countries and lead to a new international division of labour.

3. The long term

The world crisis, which began at the end of 1973, affected both the rates of mineral production and consumption and international trade in these products. However, one of the structural causes of that crisis appears to have been the exhaustion of the industrialization model of the developed countries and of the implicit international division of labour that has existed for the last three decades of which the considerable expansion of international trade is one of its fundamental bases. The annual growth rate of international trade, which during the period 1928-1938 was 1.5%, increased to 11.7% during the period 1950-1973, fell to 3.8% in the period 1973-1979 and showed negative values in 1980.

OPEC's pricing policy which was imposed in 1973, spurred the developed countries to fine-tune their policies for reducing their dependency on the imports of mining raw materials, to endeavour on the one hand, to achieve greater self-sufficiency or to diversify their external sources of supply and on the other hand, to replace the relatively scarcer products. This process of substitution was accentuated even more by the structure which derived from the greater specialization of the developed countries in high technology and service industries, whose main impact on the demand for minerals was on the one hand, to reduce the input of component of these products per unit of manufactured end use products and on the other, to generate a stronger demand for a group of "non-traditional" minerals.

The producers-exporters of minerals have usually responded to the crisis by formulating policies for building up and accumulating inventories and reducing production in order to maintain and in some cases, even raise price levels.

It may be observed that the policies of both the importing and producer-exporter countries, while they may lead to a degree of stability in the minerals market, through a contraction of demand and supply, on the other hand, they are not suitable instruments for overcoming the crisis, and initiating a new development process. In this sense, two basic assumptions could be made which might give a fresh impetus to the world mining economy.

The first assumption relates to the possibility of making significant changes in the pattern of the international division of labour, which prevailed in 1973. In the new pattern, the developing countries would specialize in the exploitation and gradual industrialization of their mining exports, from the metal-working phase to the phase of processing end use products with a high content of the heaviest metals and minerals, whereas the developed countries would base their industrial redeployment on specialization in high technology industries, with few mining resource components. One of the instruments, which could gradually give shape to this new pattern of the international division of labour in the mining sector, would be "improvement" of the long-term sales contracts which would also include, on the one hand, clauses relating to financing and transfer of technology and on the other, clauses for gradual industrialization that would involve changes in the features of the product to be marketed.

It must be borne in mind that the results of the North-South dialogue at the Paris Conference, of the Special Session of the United Nations General Assembly held in September 1980 and of the Cancún Conference on the implementation of the New International Economic Order do not favour a political decision to bring about these changes. From this situation, a second assumption of development in the mining sector would flow which, contrary to the previous one, would be a development alternative with a high degree of self-sufficiency in large regions or groups of countries.

These forms of autonomous growth would obviously require considerable specialization and national complementarity since they would need high levels of intra-regional trade in order to be dynamic. On the other hand, the degree of self-sufficiency would be dictated in each region by the volume or the supplies of mining reserves as well as by the expansion of industrial consumption of those resources. The excessive consumption over production levels would determine the extraregional import requirements, that would have to be met by exportable surpluses, generated by other regions.

On the basis of this development assumption, the industrialization of the resources of Latin America would need to have as a basic prerequisite an integrated industrial structure for inputs, which would complement end use goods. Obviously, this process of industrialization would, in turn, require financial resources and technology, which are still not available in the region and so the co-operation on the international community would be needed, perhaps in certain joint-ventures of vertically integrated production and intra-regional marketing.

Note

1/ This production could, for example, be done according to the following formula:

$$\frac{\Delta P}{P} = a + b \frac{\Delta L}{L} + c \frac{\Delta K}{K} + d \frac{\Delta N}{N} + e \frac{\Delta T}{T}$$

in which:

$\dfrac{\Delta P}{P}$ = Annual growth rates of mining production.

a = Coefficient of the variations of global productivity determined by other factors which are not included in the function indicated.

b = Coefficient of the marginal productivity of labour, whose variations would determine wage variations.

$\dfrac{\Delta L}{L}$ = Growth rate of the labour force directly employed in mining activity measured in working-hours and man-months.

c = Coefficient of the marginal productivity of capital, whose variations would determine the variations of the dividends of shareholders or financial agencies.

$\dfrac{\Delta K}{K}$ = Growth rate of the investments and working capital used directly in mining activity and measured in monetary units.

d = Coefficient of productivity of the mining deposits being worked.

$\dfrac{\Delta N}{N}$ = Growth rate of the wealth of the mineral deposit which can be measured in terms of variations in the average grade of the ore at the point of entry into the processing plant. According to the mining legislation of the countries of the region, the wealth of the subsoil, including mining deposits, is State property. The contribution of this sector to the mining output would therefore determine the level of State revenue, which is collected either by means of a tax on the variation in the wealth of the deposit or through the State's proportional participation, as an associate member of the enterprises in the profits yielded. This is the part of the economic surplus that would be earmarked for investment in new explorations, infrastructure and other forms of reproductive capital, since it is assumed that the part corresponding to labour would be almost completely consumed by the working population, and that the surplus appropriated by the capital factor would be used according to the decisions of its owners, in which case the State may only be able to promote more industrialization of the mining product through reinvestment incentives.

e = Coefficient of the productivity of technological innovations.

$\dfrac{\Delta T}{T}$ = Growth rate of the "cost-benefit" ratio of technological innovations. These costs would, for example, include the "royalties" paid on the transfer of technology and expenditure for research and training, etc. The benefits would be represented by the levels of mineralogical or metallic recovery levels in mining activities, the concentration of ores and extractive metallurgy.

Bibliography

Adam, Henri T., Evolution of the problem of guaranteeing loans for the integration projects of developing countries, UNCTAD/ST/ECDC/23(2), 27 March 1984.

Adams, F. Gerard and Behrman, Jere R., "The linkage effects of raw material processing in economic development: A survey of modeling and other approaches", Journal of policy modeling, vol. 3, No. 3, October 1981.

Association of Iron Ore Exporting Countries, Estadísticas de mineral de hierro, September 1983.

Bartoszewicz, Tomasz, "Institutional aspects of international trade in the eighties: Opportunities and barriers", Development and Peace, vol. 2, No. 1, Spring 1981.

Callot, F., World production and consumption of minerals in 1978, Mining Journal Books Ltd., 1981.

CIPEC, Quarterly Review, October-December 1983.

Demeoq, Marielle, "The rationale and modalities for compensating export earnings instability", Development and change, vol. 15, No. 3, July 1984.

Grábek Zdenek, "Efficiency in natural resources usage: A comparison of market and central planning policies", Journal of policy modeling, vol. 3, No. 1, February 1981.

ECLAC, Statistical Yearbook for Latin America and the Caribbean, 1983.

ECLAC, Aspects of a Latin American policy in the commodities sector, preliminary version, E/CEPAL/R.335/Rev.1, 27 April 1983.

ECLAC, Economic Projections Centre, Modelo macroeconómico: El desarrollo latinoamericano, su evaluación y perspectivas a largo plazo, preliminary figures as of 5 October 1984.

ECLAC, Evolution of, and prospects for, the mining sector in Latin America, E/CEPAL/R.265, May 1981.

ECLAC, Aspectos de una Política Latinoamericana en el Sector de Productos Básicos, E/CEPAL/R.335, February 1983.

95

ECLAC, La cooperación técnica y económica en el sector minero-metalúrgico de América Latina, E/CEPAL/R.331, 23 March 1984.

ECLAC, Posibilidades de cooperación y complementación industrial entre América Latina y Japón para la producción y comercialización del hierro y el acero, E/CEPAL/L.265, May 1982.

ECLAC/IPEA, Cooperación económica entre Brasil y el Grupo Andino: El caso de los minerales y metales no ferrosos, E/CEPAL/G.1268, Corr.1, January 1984.

Federal Institute for Geosciences and Natural Resources, Regional distribution of mining production and reserves of mineral commodities in the world, Hannover, January 1981.

GATT, Committees on tariff concessions and trade and development, Tariff escalation, copper producing and copper consuming industries, TAR/W/26, COM. TD/W/361, 7 April 1982

García Núñez, Luis, Problemas y experiencias en la gestión minera estatal del Perú: Período 1950-1981, June 1982.

Gilles, Malcolm, Edit. Al. Taxation and mining, Ballinger Publishing Co., Cambridge, Mass., 1978.

Inter-American Development Bank, Commodity export prospects of Latin America, March 1984.

Inter-American Development Bank, Necesidades de inversiones y financiamiento para energía y minerales en América Latina, June 1981.

International Iron and Steel Committee on Statistics, Steel statistics yearbook 1983, ISSN 0771-2871, Brussels 1984.

London Metals Research Unit Shearson Lehman/American Express Inc., Mid-year review of the metal markets 1984.

London Metals Research Unit Shearson Lehman/American Express Inc., Weekly Review, 24 July 1984.

Maizels, Alfred, "A conceptual framework for analysis of primary commodity markets", World development, vol. 12, No. 1, January 1984.

Meadows, Donella H., Dennis L. Meadows, Jorgen Randers and William W. Behrens III, The Limits to Growth. A report for the Club of Rome's project on the predicament of mankind. New York, Universe Books, 1972.

Metallgesellschaft Aktiengesellschaft, Metal statistics, 1972-1982, Frankfurt Am Main, 1983.

Metal Bulletin Journals Ltd., Metal bulletin, 19 April 1984 and other issues.

Mesarovic, Mihajlo and E. Pestel, Mankind at the turning point. The second report to the Club of Rome, New York, E.P. Dutton, 1974.

Mining Journal Ltd., Mining annual review 1984.

Neller, David, Natural resource tax policy in developing countries, International Monetary Fund, DM/84/14, 2 March 1984.

Palmer, Keith F., "Mineral taxation policies in developing countries, an application of resource rent tax", International Monetary Fund Staff Papers, vol. 27, No. 3, September 1980.

Perloff, Harvey, Joint author, Resources for the Future, Johns Hopkins Press for Resources for the Future Inc., Baltimore, 1963.

Portney, Paul R., Edit., "Current issues in natural resources policy", Johns Hopkins University Press, Baltimore, 1982.

Radetzki, Marian, "Changing structures in the financing of the minerals industry in LDCs", Development and change, vol. 11, No. 1, January 1980.

Radetzki, Marian, "Long-run price prospects for aluminium and copper", Natural Resources Forum, vol. 1, No. 1, January 1983.

Rodrik, Dani, "Managing resource dependency: The United States and Japan in the market for copper, iron ore and bauxite", World Development, vol. 10, No. 7, July 1982.

Rose, Allen J., Joint author, APL, an Interactive Approach, John Wiley and Sons, New York, 1975.

Salas, Guillermo P., Preliminary study on mineral resources of Latin America, Mexico, 1979.

Sánchez Albavera, Fernando, Cambios y tendencias en las negociaciones mineras: La experiencia peruana, April 1984.

Sánchez Albavera, Fernando, Profile and possibility of a multinational oil marketing enterprise: the case of Latin America (LC/R.385/Sem.19/10), Santiago, Chile, ECLAC, 1984.

Sánchez Albavera, Fernando, Productos básicos: Segmentación y sincronización transnacionales, E/CEPAL/R.359, 16 May 1984.

Saulniers, Alfred, Public enterprises in Latin America: The new look?, E/CEPAL/BRAS/Sem.2/R.1, 1 May 1983.

Schultze, Charles S., Joint editor, Higher oil prices and the World Economy, The Brookings Institution, Washington, D.C., 1975.

UNCTAD, Issues related to the promotion and financing of integration projects of developing countries, UNCTAD/ST/ECDC/23(3), 17 May 1984.

UNCTAD, Cuestiones que plantean el fomento y la financiación de proyectos de integración de los países en desarrollo, UNCTAD/ST/ECDC/23(3), 17 May 1984.

UNCTAD, Monthly Commodity Price Bulletin, vol. IV, No. 5, May 1984.

UNCTAD, Trade and Development Report, 1983, Part I: "The current world economic crisis", UNCTAD/TDR/3 (part I), 7 September 1983.

UNCTAD, Trade and Development Report, 1983, Part II: "Economic co-operation among developing countries", UNCTAD/TDR/3 (part II), 5 October 1983.

UNCTAD, A strategy for the technological transformation of the developing countries, TD/277, June 1983.

UNDTCD and DSE of German Foundation for International Development, Legal and institutional arrangements in minerals development, Mining Journal Books, 1982.

UNIDO, Mineral processing in developing countries, ID/253, New York, 1980. United Nations publication, Sales No.: 80.II.B.5.

United Nations, Centre on Transnational Corporations, Salient features and trends in foreign direct investment, ST/CTC/14, New York, 1983. United Nations Publication, Sales No.: 83.II.A.8.

United Nations, Centre on Transnational Corporations, <u>Main features and trends in petroleum and mining agreements</u>, ST/CTC/29, New York, 1983.

United Nations, Centre on Transnational Corporations, <u>National legislation and regulations relating to transnational corporations: A technical paper</u>, ST/CTC/35, New York, 1983.

United Nations, Economic and Social Council, <u>Regional Co-operation</u>, E/1984/112, 12 June 1984.

United Nations, <u>Monthly bulletin of statistics</u>, vol. XXXVIII, No. 5, May 1984.

United Nations, <u>World population prospects as assessed in 1980</u>, ST/ESA/SER.A/78, New York 1981.

United States Department of the Interior, Bureau of Mines, <u>Mineral facts and problems</u>, Commodity data summaries, 1980.

United States Department of the Interior, Bureau of Mines, <u>Minerals Yearbook</u>, vol. I, several issues.

Walde, Thomas W., "Permanent sovereignty over natural resources, Recent developments in the mineral sector", <u>Natural Resources Forum</u>, vol. 7, No. 3, July 1983.

Walde, Thomas W., "Third world mineral development: Current issues", <u>The Columbia Journal of World Business</u>, vol. XIX, No. 1, Spring 1984.

World Bank, Commodity trade and price trends, 1983-1984 edition.

World Bank, Office Memorandum, <u>Half-yearly revision of commodity price forecasts and quarterly review of commodity markets</u>, 20 January 1982.

World Bank, Office Memorandum, <u>Primary commodity price forecasts</u>, 16 July 1984.

World Bank, <u>Price prospects for major primary commodities</u>, Reports No. 814/82, July 1982.

World Bureau of Metal Statistics, <u>World Metal Statistics</u>, vol. 37, No. 40, April 1984.

World Bureau of Metal Statistics, <u>World Metal Statistics Yearbook</u>, 1984.

STATISTICAL APPENDIX

Table 1

EVOLUTION OF WORLD PRODUCTION OF THE MAIN MINERALS
(In thousands of metric tons of high grade ore)

Minerals	1947	1960	1965	1974	1975
Antimony	38	-	63	72	68
Asbestos	838	2 205	3 570	4 589	4 509
Barite	1 516	-	3 438	4 485	4 804
Bauxite	6 154	27 620	36 530	76 810	73 939
Bismuth a/	1 509	-	4 264	4 826	3 578
Cadmium	5 509	-	27 800	19 038	16 906
Chromium	1 842	4 401	4 899	7 427	7 930
Cobalt a/	7 085	14 031	15 422	32 469	32 914
Columbium a/	-	-	6 350	13 258	12 757
Copper	2 274	4 402	5 600	8 063	7 679
Fluorite	687	1 807	2 876	4 842	4 621
Gold	726	-	1 644	1 377	1 330
Ilmenite	242	1 102	2 475	2 811	2 589
Iron Ore	144 713	242 231	617 997	902 870	897 800
Lead	1 409	2 376	2 975	3 858	3 714
Lithium	-	-	-	7	6
Magnesium	18	-	97	132	129
Manganese	4 153	13 610	17 605	22 741	24 400
Mercury a/	5 377	-	9 479	9 203	4 481
Metallic Arsenic	58	-	62	51	46
Metallic Tellurium a/	-	-	153	203	149
Metallic Selenium a/	-	-	789	1 229	1 138
Molybdenum	13	40	52	86	81
Nickel	131	342	428	790	816
Phosphated rocks	18 226	41 860	26 440	122 147	118 586
Platinum	16	39	102	199	199
Potassium	2 580	9 082	14 800	26 432	27 423
Rutile	30	100	220	331	351
Silver	5 649	-	8 652	10 166	10 143
Sulphur	3 866	-	15 120	22 271	22 119
Tantalum a/	-	-	399	412	408
Thorium	-	-	-	12	12
Tin	114	190	199	234	225
Tungsten	14	31	17	37	37
Vanadium	1	7	8	19	22
Zinc	1 817	3 351	4 750	6 281	6 131

(cont.Table 1)

Minerals	1978	1980	1981	1982	1983
Antimony	62	64	57	54	48
Asbestos	4 693	4 902	4 480	4 311	4 176
Barite	6 885	7 578	8 216	7 155	-
Bauxite	80 975	92 623	85 474	74 441	78 568
Bismuth a/	4 254	3 421	3 382	3 248	4 153
Cadmium	17 468	18 663	17 535	16 755	17 664
Chromium	10 944	9 729	10 647	9 895	7 880
Cobalt a/	26 823	32 724	30 274	25 084	24 127
Columbium a/	9 666	-	14 816	14 316	-
Copper	7 604	7 816	8 175	7 963	8 220
Fluorite	4 665	4 682	5 051	4 539	4 301
Gold	1 346	1 208	1 421	1 472	1 419
Ilmenite	3 515	2 914	3 638	3 058	-
Iron Ore	840 340	508 976	852 210	779 270	750 542
Lead	3 372	3 603	3 343	3 451	3 451
Lithium	2	6	2	2	81
Magnesium	288	321	296	248	-
Manganese	22 642	26 697	23 543	22 456	21 992
Mercury a/	6 239	6 622	7 377	7 032	5 668
Metallic Arsenic	31	-	28	26	-
Metallic Tellurium a/	152	508	105	97	-
Metallic Selenium a/	1 442	1 905	1 302	1 217	-
Molybdenum	100	108	109	91	63
Nickel	659	748	712	608	655
Phosphated rocks	125 022	136 014	137 524	122 633	-
Platinum	222	212	239	222	202
Potassium	26 122	27 871	27 046	26 230	-
Rutile	302	447	371	346	-
Silver	11 891	10 422	12 489	12 841	12 393
Sulphur	52 138	-	53 563	50 660	-
Tantalum a/	362	-	371	335	-
Thorium	22	-	20	20	-
Tin	241	235	253	241	210
Tungsten	46	53	49	45	41
Vanadium	31	34	35	33	-
Zinc	5 854	6 247	5 657	6 047	6 498

CHANGES IN THE STRUCTURE OF THE VALUE OF WORLD MIN. 11.

Minerals	Annual growth rates		
	1965-1974	1974-1980	1980-1983
Antimony	1.5	- 1.9	- 9.1
Asbestos	12.8	1.1	-5.2
Barite	3.0	9.1	-
Bauxite	8.6	3.2	-5.3
Bismuth a/	1.4	-5.6	6.7
Cadmium	-4.1	-0.3	-1.8
Chromium	4.7	4.6	-6.8
Cobalt a/	8.6	0.1	-9.7
Columbium a/	8.5	-	.
Copper	4.1	-0.5	1.7
Fluorite	6.0	-0.6	-2.8
Gold	-1.9	-2.2	5.5
Ilmenite	1.4	0.6	.
Iron Ore	4.3	-9.1	13.8
Lead	2.9	-1.3	-1.4
Lithium	-	-2.5	138.1
Magnesium	3.5	16.0	.
Manganese	2.9	2.7	-6.3
Mercury a/	-0.3	-5.3	-5.1
Metallic Arsenic	-2.1	-	.
Metallic Tellurium a/	3.2	16.5	.
Metallic Selenium a/	5.0	7.6	.
Molybdenum	3.0	3.9	-16.4
Nickel	7.1	-0.9	-4.3
Phosphated rocks	18.5	1.8	.
Platinum	7.7	1.1	-1.6
Potassium	6.7	0.9	.
Rutile	4.6	5.1	.
Silver	1.8	0.4	5.9
Sulphur	4.4	-	.
Tantalum a/	0.4	-	.
Thorium	-	-	.
Tin	1.8	0.1	-3.7
Tungsten	9.0	6.2	-8.2
Vanadium	10.1	-	.
Zinc	3.1	-0.1	1.3

Sources: 1. U.S. Department of the Interior "Minerals Yearbook
Vol.I Metals and Minerals". Various issues.
2. World Bureau of Metal Statistics "World Metal
Statistics" Vol. 37 N° 40, April,1984.
3. Association of Iron Ore Exporting Countries,
"Iron Ore Statistics", September 1983.

a/ Tons.

Table 2

CHANGES IN THE STRUCTURE OF THE VALUE OF WORLD MINERAL PRODUCTION
(Percentages)

1950		1973		1978	
Iron Ore	18.5	Copper	24.9	Iron Ore	16.2
Copper	15.3	Iron Ore	17.5	Copper	12.0
Gold	15.2	Gold	10.4	Gold	10.5
Lead	6.2	Zinc	3.8	Phosphates	4.2
Zinc	5.9	Nickel	3.2	Uranium	3.7
Manganese	3.0	Lead	2.9	Potassium	3.5
Potassium	2.4	Potassium	2.8	Diamonds	2.8
Silver	2.0	Diamonds	2.7	Lead	2.8
Phosphates	2.0	Phosphates	2.2	Zinc	2.7
Sulphur	1.7	Silver	1.9	Amianto	2.7
Amianto	1.5	Amianto	1.4	Silver	2.6
Nickel	1.4	Platinum	1.4	Bauxite	2.2
Industrial diamonds	1.4	Bauxite	1.4	Nickel	2.1
Nitrates	1.0	Uranium	1.3	Sulphur	1.8
Cadmium and kaolin	0.9	Sulphur	1.2	Platinum	1.6
Bauxite	0.8	Manganese	1.1	Molybdenum	1.4
Chromium	0.8	Kaolin	0.9	Manganese	1.3
Tungsten	0.7	Molybdenum	0.7	Kaolin	1.1
Mica	0.6	Fluorite	0.6	Tungsten	1.1
Platinum	0.5	Tungsten	0.5	Chromium	0.9
Fluourite	0.5	Chromium	0.4	Borate	0.6
Antimony	0.5	Talc	0.4	Talc	0.6
Molybdenum	0.4	Borate	0.4	Magnesium	0.5
Bentonite	0.4	Vanadium	0.3	Sodium	0.5
Talc	0.4	Sodium	0.3	Fluorite	0.5
Magnesium	0.3	Magnesium	0.2	Vanadium	0.5
Barite	0.3	Bentonite	0.2	Cobalt	0.4
Borate	0.2	Barite	0.2	Bentonite	0.4
Cobalt	0.2	Ilmenite	0.2	Barite	0.3
Mercury	0.2	Antimony	0.2	Ilmenite	0.2
Ilmenite	0.2	Cobalt	0.2	Antimony	0.2
Asphalts	0.2	Mercury	0.2	Mica	0.2
Sodium	0.1	Mica	0.1	Feldspar	0.1
Graphite	0.1	Rutile	0.1	Rutile	0.1
Feldspar	0.1	Feldespar	0.1	Graphite	0.1
Rutile	-	Nitrate	0.1	Nitrate	0.1
Beryllium	-	Columbium	0.1	Columbium	0.1
Vanadium	-	Tantalum	0.1	Tantalum	-
Others	14.1	Graphite	0.1	Asphalts	-
		Asphalts	-	Mercury	-
		Others	13.3	Others	17.4
Total	100.0		100.0		100.0
Total of U.S. dollars in millions	16 919.8		59 235.6		71 492.7

(Conclusion Table 2)

	Annual growth rate	
	1950-1973	1973-1978
Iron ore	5.3	2.2
Copper	7.9	- 10.2
Gold	3.9	4.2
Phosphates	5.9	19.8
Uranium	-	28.4
Potassium	6.2	8.7
Diamonds	8.7	4.8
Lead	2.2	3.1
Zinc	3.7	- 3.3
Amianto	5.3	17.5
Silver	5.4	10.1
Bauxite	8.1	14.3
Nickel	9.6	- 4.2
Sulphur	4.0	13.2
Platinum	10.5	6.6
Molybdenum	7.9	18.0
Manganese	0.8	7.6
Kaolin	5.8	7.9
Tungsten	4.1	22.7
Chromium	2.1	23.1
Borate	7.2	16.2
Talc	5.8	12.0
Magnesium	5.2	22.6
Sodium	8.2	20.0
Flourite	6.3	- 0.2
Vanadium	18.0	2.5
Cobalt	4.9	24.8
Bentonite	3.3	14.9
Barite	4.6	12.8
Ilmenite	6.0	0.6
Antimony	1.6	- 1.4
Mica	- 0.9	4.6
Feldspar	7.6	3.6
Rutile	11.9	- 2.9
Graphite	2.3	12.8
Nitrate	- 4.5	- 5.9
Columbium	-	5.6
Tantalum	-	5.0
Asphalts	- 1.0	11.1
Mercury	5.4	- 23.7
Others	3.8	10.5
Total	3.9	7.9

Source: Callot, F., "World production and consumption of
minerals in 1978", Mining Journal Books Ltd., 1981.

Table 3

RELATIVE EVOLUTION OF WORLD MINERALS PRODUCTION
(In units of metric tons (MT))

Minerals	Unit	Latin America			Other developing countries			Developed countries		
		1960	1980	1983	1960	1980	1983	1960	1980	1983
Asbestos	Thousands MT	43	140	136	140	393	286	1 341	1 968	1 393
Bauxite	Thousands MT	13 181	24 171	16 986	2 886	19 616	17 220	6 425	37 086	32 825
Chromium	Thousands MT	125	350	160	1 949	2 138	1 656	1 072	3 665	2 656
Cobalt	MT	62	1 724	118	11 278	22 766	15 695	1 631	4 470	4 305
Copper	Thousands MT	716	1 433	1 821	1 132	1 913	1 877	1 557	2 659	2 513
Fluorite	Thousands MT	363	920	689	35	563	459	826	1 574	1 188
Ilmenite	Thousands MT	-	-	-	237	236	-	849	2 444	-
Iron Ore	Thousands MT	20 024	70 885	120 509	27 878	70 318	73 105	123 358	192 694	224 858
Lead	Thousands MT	350	370	470	305	315	247	1 132	1 861	1 745
Manganese	Thousands MT	1 099	2 824	1 935	3 323	4 127	3 652	1 776	7 779	4 365
Molybdenum	MT	1 912	14 339	23 860	72	463	277	31 927	80 666	24 443
Nickel	Thousands MT	15	60	47	40	157	128	211	326	241
Phosphated rocks	Thousands MT	910	3 021	-	14 514	42 364	-	18 145	57 922	-
Platinum	Kilograms	30	404	622	969	4	128	28 472	110 945	88 896
Potassium	Thousands MT	15	5	-	83	813	-	6 218	15 619	-
Rutile	MT	205	450	-	1 900	79 573	-	97 345	351 748	-
Tin	Thousands MT	22	35	41	109	147	113	7	17	18
Tungsten	MT	1 829	4 359	5 073	4 730	8 100	5 366	6 982	14 858	9 223
Uranium	MT	45	187	-	400	9 175	-	18 166	34 603	-
Vanadium	MT	3	455	-	890	-	-	5 672	19 725	-
Zinc	Thousands MT	528	919	1 044	257	327	360	1 810	3 357	3 468

Source: Callot, F., World production and consumption of minerals in 1983, Mining Journal Books Ltd., 1984.

Minerals	Unit	Centrally-planned economy countries			World production		
		1960	1980	1983	1960	1980	1983
Asbestos	Thousands MT	681	2 401	2 361	2 205	4 902	4 176
Bauxite	Thousands MT	5 128	11 750	11 537	27 620	92 623	78 568
Chromium	Thousands MT	1 255	3 576	3 408	4 401	9 729	7 880
Cobalt	MT	1 060	3 765	4 010	14 031	32 724	24 128
Copper	Thousands MT	637	1 812	2 009	4 042	7 817	8 220
Fluorite	Thousands MT	583	1 625	1 965	1 807	4 682	4 301
Ilmenite	Thousands MT	16	235	-	1 102	2 915	-
Iron Ore	Thousands MT	70 971	175 080	332 070	242 231	508 977	750 542
Lead	Thousands MT	589	1 057	989	2 376	3 603	3 451
Manganese	Thousands MT	7 412	11 967	12 040	13 610	26 697	21 992
Molybdenum	MT	6 487	12 533	14 250	40 398	108 001	62 830
Nickel	Thousands MT	76	205	239	342	748	655
Phosphated rocks	Thousands MT	8 291	32 707	-	41 860	136 014	-
Platinum	Kilograms	10 356	101 086	111 973	39 827	212 439	201 619
Potassium	Thousands MT	2 766	11 434	-	9 082	27 871	-
Rutile	MT	100	15 000	-	99 550	446 771	-
Tin	Thousands MT	52	36	38	190	235	210
Tungsten	MT	17 654	26 002	21 080	31 195	53 319	40 742
Uranium	MT	...	13 780	-	18 611	43 965	-
Vanadium	MT	100	...	-	6 665	33 960	-
Zinc	Thousands MT	756	1 645	1 626	3 351	6 248	6 498

Source: Federal Institute for Geosciences and Natural Resources, "Regional Distribution of Mining Production and Reserves of Mineral Commodities in the World", Hannover, January 1982.

Table 4

RELATIVE EVOLUTION OF MINING PRODUCTION 1960-1980
(<u>Annual growth rates of the volume of production</u>)
(Percentages)

Main Minerals	Latin America	Other developing countries	Developed countries	Centrally-planned economy countries	World total	Percentage of the share of the main producers in the levels of Latin America of 1980
Asbestos	6.08	5.30	1.94	6.50	4.08	Brazil(100)
Bauxite	3.08	10.06	9.16	4.23	6.24	Jamaica(48), Suriname(19), Brazil(17), Guyana(12)
Chromium	5.28	0.46	6.34	5.38	4.05	Brazil(91)
Cobalt	18.09	3.57	5.17	6.54	4.33	Cuba(100)
Copper	3.53	2.66	2.71	5.37	3.35	Chile(66), Peru(23), Mexico(11)
Fluorite	4.76	14.90	3.28	5.26	4.88	Mexico(99), Argentina(1)
Ilmenite	-	-0.02	5.43	14.38	4.98	-
Iron Ore	6.52	4.73	2.26	4.62	3.78	Brazil(74), Venezuela(11), Chile(5), Mexico(5), Peru(4)
Lead	0.28	0.16	2.52	2.97	2.10	Peru(40), Mexico(39), Argentina(9), Brazil(7), Bolivia(5)
Manganese	4.83	2.19	7.66	2.42	3.43	Brazil(78), Mexico(16), Argentina(2), Bolivia(1)
Molybdenum	10.60	9.75	4.74	3.35	5.00	Chile (93), Peru (7)

Main Minerals	Latin America	Other developing countries	Developed countries	Centrally-planned economy countries	World total	Percentage of the share of the main producers in the levels of Latin America of 1980
Nickel	7.18	7.08	2.20	5.09	3.99	Cuba(57), Dominican Republic(24), Guatemala(9), Brazil(3)
Phosphated rocks	6.18	5.50	5.98	7.10	6.07	Brazil(91), Mexico(9)
Platinum	13.88	-24.00	7.04	12.07	8.73	Colombia(100)
Potassium	-5.34	12.09	4.71	7.35	5.77	Chile(100)
Rutile	4.01	20.53	6.63	28.47	7.80	Brazil(100)
Tin	2.35	1.51	4.54	-1.82	1.07	Bolivia(75), Brazil(19), Argentina(3), Peru(3)
Tungsten	4.44	2.73	3.85	1.95	2.72	Bolivia(62), Brazil(22), Peru(10), Mexico(5), Argentina(1)
Uranium	7.38	16.96	3.27	-	4.39	Argentina(100)
Vanadium	28.54	-	6.43	27.93	8.48	Chile(100)
Zinc	2.81	1.21	3.14	3.96	3.16	Peru(57), Mexico(26), Brazil(8), Bolivia(6), Argentina(3)

Source: See Tables 1 and 3 of the Statistical Appendix.

Table 5

LATIN AMERICA: ESTIMATES OF MINERALS PRODUCTION IN 1980

Minerals	Units	World production	Latin America
Antimony	MT a/	64 635	19 326
Asbestos	Thousands MT	4 902	140
Barite	Thousands MT	7 578	1 175
Bauxite	Thousands MT	92 623	25 193
Bismuth	MT	3 421	1 279
Cadmium	MT	18 663	952
Chromium	Thousands MT	9 729	350
Cobalt	MT	32 724	1 724
Copper	Thousands MT	7 816	1 610
Fluorite	Thousands MT	4 682	920
Gold	MT	1 208	79
Iron Ore	Thousands MT	508 976	93 143
Lead	Thousands MT	3 603	370
Lithium	MT	6 615	3 903
Magnesium	Thousands MT	321	-
Manganese	Thousands MT	26 697	2 824
Mercury	MT	6 622	73
Molybdenum	Thousands MT	108	14
Nickel	Thousands MT	748	67
Niobium	Thousands MT	15	13
Phosphated rocks	Thousands MT	136 014	3 021
Platinum	MT	212	0.4
Potassium	Thousands MT	27 871	28
Rutile	Thousands MT	447	0.4
Selenium	MT	1 905	318
Silver	MT	10 421	3 366
Tellurium	MT	508	95
Tin	Thousands MT	235	36
Tungsten	MT	53 340	5 438
Uranium	MT	43 965	187
Vanadium	MT	33 960	455
Zinc	Thousands MT	6 248	919

(Cont.Table 5)

Minerals	Unit	Argentina	Bolivia	Brazil
Antimony	MT	-	15 448	65
Asbestos	Thousands MT	-	-	140
Barite	Thousands MT	61	8	106
Bauxite	Thousands MT	-	-	4 168
Bismuth	MT	-	10	-
Cadmium	MT	19	-	37
Chromium	Thousands MT	-	-	320
Cobalt	MT	-	-	-
Copper	Thousands MT	-	-	-
Fluorite	Thousands MT	2	-	-
Gold	MT	-	1	40
Iron Ore	Thousands MT	510	-	68 712
Lead	Thousands MT	32	18	25
Lithium	MT	26	-	53
Magnesium	Thousands MT	-	-	-
Manganese	Thousands MT	56	30	2 189
Mercury	MT	-	-	-
Molybdenum	Thousands MT	-	-	-
Nickel	Thousands MT	-	-	2
Niobium	Thousands MT	-	-	13
Phosphated rocks	Thousands MT	-	-	2 749
Platinum	MT	-	-	-
Potassium	Thousands MT	-	-	-
Rutile	Thousands MT	-	-	0.4
Selenium	MT	-	-	-
Silver	MT	73	188	-
Tellurium	MT	-	-	-
Tin	Thousands MT	1	27	7
Tungsten	MT	53	3 359	1 226
Uranium	MT	187	-	-
Vanadium	MT	-	-	-
Zinc	Thousands MT	31	51	69

(Cont.Table 5)

Minerals	Unit	Chile	Colombia	Mexico
Antimony	MT	-	-	2 198
Asbestos	Thousands MT	-	-	-
Barite	Thousands MT	205	-	333
Bauxite	Thousands MT	-	-	4 168
Bismuth	MT	-	-	749
Cadmium	MT	-	168	719
Chromium	Thousands MT	-	-	-
Cobalt	MT	-	-	-
Copper	Thousands MT	1 071	-	172
Fluorite	Thousands MT	-	-	918
Gold	MT	4	8	6
Iron Ore	Thousands MT	5 090	-	5 088
Lead	Thousands MT	-	-	144
Lithium	MT	3 824	-	-
Magnesium	Thousands MT	-	-	-
Manganese	Thousands MT	27	-	454
Mercury	MT	-	-	53
Molybdenum	Thousands MT	13	-	-
Nickel	Thousands MT	-	-	-
Niobium	Thousands MT	-	-	-
Phosphated rocks	Thousands MT	-	-	272
Platinum	MT	-	0.4	-
Potassium	Thousands MT	28	-	-
Rutile	Thousands MT	-	-	-
Selenium	MT	223	-	19
Silver	MT	302	-	1 469
Tellurium	MT	67	-	5
Tin	Thousands MT	-	-	-
Tungsten	MT	-	-	267
Uranium	MT	-	-	-
Vanadium	MT	455	-	-
Zinc	Thousands MT	-	-	237

(Conclusion Table 5)

Minerals	Units	Peru	Venezuela	Other Countries
Antimony	MT	776	-	839
Asbestos	Thousands MT	-	-	-
Barite	Thousands MT	71	-	391
Bauxite	Thousands MT	-	-	21 025
Bismuth	MT	520	-	-
Cadmium	MT	9	-	-
Chromium	Thousands MT	-	-	30
Cobalt	MT	-	-	1 724
Copper	Thousands MT	367	-	-
Fluorite	Thousands MT	-	-	-
Gold	MT	5	-	15
Iron Ore	Thousands MT	3 563	10 180	-
Lead	Thousands MT	151	-	-
Lithium	MT	-	-	-
Magnesium	Thousands MT	-	-	-
Manganese	Thousands MT	-	-	68
Mercury	MT	-	-	20
Molybdenum	Thousands MT	1	-	-
Nickel	Thousands MT	-	-	65
Niobium	Thousands MT	-	-	-
Phosphated rocks	Thousands MT	-	-	-
Platinum	MT	-	-	-
Potassium	Thousands MT	-	-	-
Rutile	Thousands MT	-	-	-
Selenium	MT	76	-	-
Silver	MT	1 230	-	104
Tellurium	MT	23	-	-
Tin	Thousands MT	1	-	-
Tungsten	MT	533	-	-
Uranium	MT	-	-	-
Vanadium	MT	-	-	-
Zinc	Thousands MT	531	-	-

Source: See Table 1 of the Statistical Appendix.

a/ MT = metric tons

Table 6

WORLD MINERAL RESERVES IN 1981
(In units of metric tons (MT))

Units	Minerals	Latin America	Other developing countries	Developed countries	Centrally planned economy countries	Total	Percentage of reserves in the five major producing countries	
							%	Countries
Thousands MT	Amianto	...	...	...	...	87 000	73%	(USSR 32, Canada 25, S.Africa 6, Rhodesia 6, U.S.A. 4)
Thousands MT	Antimony	650	320	840	2 495	4 305	78%	(China 51, Bolivia 9, S.Africa 7, USSR 6, Mexico 5)
Thousands MT	Asbestos	5 540	8 360	66 000	43 200	123 100	85%	(Canada 37, USSR 33, S.Africa 7, Zimbabwe 4, Brazil 4)
Thousands MT	Barite	15 378	65 922	108 700	43 000	233 000	54%	(U.S.A. 22, India 12, China 9, Canada 6, USSR 5)
Millions MT	Bauxite	6 131	10 679	5 790	800	23 400	74%	(Guinea 28, Australia 20, Brazil 11, Jamaica 9, India 6)
MT	Bismuth	23 655	5 495	58 550	7 300	101 660	72%	(Japan 24, Australia 18, Bolivia 14, U.S.A. 10, Mexico 6)
Thousands MT	Cadmium	69	116	440	55	680	60%	(Canada 18, U.S.A. 16, Australia 14, USSR 7, Ireland 5)
Millions MT	Chromium	7	1 024	2 302	208	3 541	99%	(South Africa 64, Zimbabwe 28, USSR 6, Finland 1)
Thousands MT	Cobalt	44	2 246	375	1 000	3 665	72%	(Cuba 22, Indonesia 15, Zaire 12, Philippines 12, New Caledonia 11)
Thousands MT	Columbium	8 165	...	...	...	10 600	96%	(Brazil 77, USSR 6, Canada 6, Zaire 4, Uganda 3)

Units	Minerals	Latin America	Other developing countries	Developed countries	Centrally-planned economy countries	Total	Percentage of reserves in the five major producing countries	
							%	Countries
Thousands MT	Copper	189 445	148 654	160 500	72 000	570 599	57%	(Chile 19, U.S.A. 18, USSR 7, Zambia 7, Canada 6)
Thousands MT	Fluorite	52 419	45 381	176 200	29 000	303 000	66%	(South Africa 36, Mexico 13, U.K. 7, U.S.A. 5, Kenya 5)
MT	Gold	387	4 123	19 968	7 776	32 254	85%	(S.Africa 51, USSR 24, U.S.A. 4, Australia 4, Philippines 2)
Thousands MT	Ilmenite	800	67 662	267 150	58 500	394 112	77%	(Canada 25, Norway 19, India 13, USSR 12, S.Africa 8)
Millions MT	Iron Ore	53 773	490	33 150	31 600	119 013	93%	(USSR 30, Bolivia 22, Brazil 17, Canada 12, Australia 12)
Thousands MT	Lead	13 163	9 637	105 400	28 500	156 700	68%	(U.S.A. 27, Australia 14, Canada 13, USSR 11, S.Africa 3)
Thousands MT	Lithium	1 299	642	253	...	2 194	99%	(Chile 59, Zaire 25, Canada 9, Zimbabwe 4, U.S.A. 2)
Millions MT	Magnesium	473	1 592	668	6 457	8 763	81%	(China 29, USSR 26, Korea 18, Brazil 5, Australia 3)
Millions MT	Manganese	150	98	886	700	1 834	98%	(S.Africa 45, USSR 38, Australia 8, Gabon 5, Brazil 2)
MT	Mercury	8 584	18 316	96 000	63 700	186 600	77%	(Spain 28, USSR 20, China 13, Yugoslavia 9, U.S.A.7)

(Cont. Table 6)

Units	Minerals	Latin America	Other developing countries	Developed countries	Centrally-planned economy countries	Total	Percentage of reserves in the five major producing countries	
							%	Countries
Thousands MT	Molybdenum	3 223	419	4 883	955	9 480	85%	(U.S.A. 44, Chile 26, USSR 7, Canada 6, Panama 2)
Thousands MT	Nickel	23 879	34 306	18 120	24 600	100 905	71%	(New Caledonia 19, Cuba 18, Canada 12, USSR 11, Indonesia 11)
Thousands MT	Niobium	6 543	562	135	700	7 940	97%	(Brazil 82, USSR 9, Zaire 3, Canada 2, Nigeria 1)
Millions MT	Phosphated Rocks	1 490	48 130	11 800	9 500	70 920	87%	(Morocco 59, U.S.A. 12, USSR 11, South Africa 3, Australia 2)
MT	Platinum	31	-	30 527	6 220	36 778	100%	(S.Africa 82, USSR 16, Canada 1, Colombia 1)
Millions MT	Potassium	64	416	3 680	4 920	9 080	92%	(USSR 44, Canada 30, German Democratic Republic 9, Federal Republic of Germany 6, Israel 3)
Millions MT	Rutile	58	12 192	13 750	2 900	28 900	87%	(Australia 34, India 25, S.Leone 13, USSR 8, U.S.A. 7)
MT	Selenium	57 264	73 995	66 151	19 500	216 910	54%	(Chile 18, U.S.A. 17, USSR 7, Canada 6, Peru 6)
MT	Silver	53 055	5 150	114 510	57 900	230 675	76%	(USSR 2, U.S.A. 20, Australia 13, Mexico 12, Canada 10)
MT	Tantalum	3 625	53 536	4 220	4 530	65 911	87%	(Zaire 56, Nigeria 11, Thailand 7, USSR 7, Malyasia 6)

Units	Minerals	Latin America	Other developing countries	Developed countries	Centrally-planned economy countries	Total	Percentage of reserves in the five major producing countries	
							%	Countries
MT	Tellurium	3 200	33 575	18 109	5 500	60 384	32%	(U.S.A. 15, USSR 6, Canada 5, Peru 5, Japan 1)
Thousands MT	Tin	1 587	5 066	730	2 520	9 903	65%	(Indonesia 16, China 15, Thailand 12, Malysia 12, USSR 10)
Thousands MT	Tungsten	111	240	595	1 689	2 635	79%	(China 52, Canada 10, USSR 8, U.S.A. 5, North Korea 4)
Thousands MT	Uranium	225	307	2 059	...	2 591	75%	(U.S.A. 27, S. Africa 15, Sweden 12, Australia 12, Canada 9)
Thousands MT	Vanadium	223	97	8 220	7 395	15 935	98%	(S.Africa 49, USSR 46, Australia 1, Chile 1, China 1)
Thousands MT	Zinc	15 907	14 193	175 920	35 000	241 020	68%	(Canada 26, U.S.A. 20, Australia 10, USSR 8, Japan 4)
Thousands MT	Zirconium	895	13 095	25 350	5 400	44 740	92%	(Australia 32, India 25, U.S.A. 16, USSR 11, South Africa 8)

Source: 1. Federal Institute for Geosciences and Natural Resources, "Regional Distribution of Mining Production and Reserves of Mineral Commodities in the World", Hannover, January 1982.
2. U.S. Bureau of Mines "Minerals Facts and Problems/Commodity Data Summaries, 1980".

Table 7

WORLD MINERAL RESERVES IN 1983
(In units of metric tons (MT))

Unit	Minerals	Latin America		Other developing countries	
		Quantity	%	Quantity	%
Thousands MT	Antimony	650	15.1	320	7.4
Thousands MT	Asbestos	5 650	4.6	8 250	6.7
Millions MT	Bauxite	6 270	26.8	10 540	45.0
MT	Bismuth	23 700	24.9	5 450	5.7
Millions MT	Chromium	7	0.2	1 024	28.9
Thousands MT	Cobalt	45	1.2	2 245	61.3
Thousands MT	Copper	194 500	35.3	123 800	22.5
Thousands MT	Fluorite	52 400	17.3	45 400	14.9
MT	Gold	1 886	5.8	2 644	8.2
Millions MT	Iron Ore	19 230	20.5	9 620	10.3
Thousands MT	Lead	14 100	9.0	8 700	5.5
Thousands MT	Lithium	1 299	59.2	642	29.3
Millions MT	Manganese	42	2.3	129	7.0
MT	Mercury	9 600	5.1	17 300	9.3
Thousands MT	Molybdenum	3 225	34.0	418	4.4
Thousands MT	Niobium	6 545	82.4	560	7.1
Thousands MT	Nickel	5 000	6.1	34 310	41.8
MT	Platinum	31	0.1	-	-
MT	Silver	53 150	23.0	5 055	2.2
Thousands MT	Tin	1 395	14.4	5 070	52.2
Thousands MT	Tungsten	108	4.1	243	9.2
Thousands MT	Zinc	18 100	7.5	12 000	5.0

(Conclusion Table 7)

Unit	Minerals	Developed countries Quantity	%	Centrally-planned countries Quantity	%	World Total Quantity
Thousands MT	Antimony	840	19.5	2 495	58.0	4 305
Thousands MT	Asbestos	66 000	53.6	43 200	35.1	123 100
Millions MT	Bauxite	5 790	24.7	800	3.4	23 400
MT	Bismuth	58 550	61.6	7 300	7.7	95 000
Millions MT	Chromium	2 302	65.0	208	5.9	3 541
Thousands MT	Cobalt	375	10.2	1 000	27.3	3 665
Thousands MT	Copper	160 500	29.1	72 000	13.1	550 800
Thousands MT	Fluorite	176 200	58.2	29 000	9.6	303 000
MT	Gold	19 968	61.9	7 776	24.1	32 254
Millions MT	Iron Ore	33 150	35.4	31 600	33.8	93 600
Thousands MT	Lead	105 400	67.3	28 500	18.2	156 700
Thousands MT	Lithium	253	11.5	-	-	2 194
Millions MT	Manganese	963	52.5	700	38.2	1 835
MT	Mercury	96 000	51.4	63 700	34.1	186 600
Thousands MT	Molybdenum	4 882	51.5	955	10.1	9 480
Thousands MT	Niobium	135	1.7	700	8.8	7 940
Thousands MT	Nickel	18 120	22.1	24 600	30.0	82 030
MT	Platinum	30 527	83.0	6 220	16.9	36 778
MT	Silver	114 510	49.6	57 960	25.1	230 675
Thousands MT	Tin	730	7.5	2 520	25.9	9 715
Thousands MT	Tungsten	595	22.6	1 689	64.1	2 635
Thousands MT	Zinc	175 920	73.0	35 000	14.5	241 020

Source: See Table 6 of the Statistical Appendix.

Table 8

EVOLUTION OF CONSUMPTION OF THE MAIN MINERALS

	1965	1974	1980	1981	1982	1983
1. Bauxite (Thousands of MT)						
a) Latin America	...	2 402	3 637	3 235	2 675	2 573
b) Other developing countries	...	2 630	4 774	4 937	4 637	5 104
c) Developed countries	...	62 432	63 782	59 316	51 503	55 712
d) Centrally-planned economy countries	...	16 787	21 133	21 225	19 366	...
Total	...	84 251	93 326	88 713	78 181	...
2. Cadmium (MT)						
a) Latin America	44	273	576	605	581	717
b) Other developing countries	82	190	462	428	608	610
c) Developed countries	10 256	13 328	12 083	11 831	11 594	11 800
d) Centrally-planned economy countries	1 990	3 200	3 861	3 628	3 690	-
Total	12 372	16 991	16 982	16 492	16 473	-
3. Copper (Thousands of MT)						
a) Latin America	167	315	497	434	457	324
b) Other developing countries	88	166	337	407	394	466
c) Developed countries	4 774	6 014	6 165	6 269	5 765	5 831
d) Centrally-planned economy countries	1 165	1 845	2 423	2 415	2 448	-
Total	6 193	8 340	9 422	9 525	9 064	...
4. Tin (Thousands of MT)						
a) Latin America	6	9	11	8	10	10
b) Other developing countries	11	11	13	13	12	14
c) Developed countries	153	172	142	134	123	130
d) Centrally-planned economy countries	40	52	56	54	55	-
Total	210	244	222	209	200	-
5. Iron ore (Millions of MT)						
a) Latin America	10	19	16	14	23	-
b) Other developing countries	13	15	27	23	32	-
c) Developed countries	193	301	267	250	206	-
d) Centrally-planned economy countries	111	161	198	190	190	-
Total	327	496	508	477	451	-

(Cont. Table 8)

	1965	1974	1980	1981	1982	1983
6. Magnesium (Thousands of MT)						
a) Latin America	2	15	20	9	10	-
b) Other developing countries	-	1	4	4	5	-
c) Developed countries	132	206	180	167	152	-
d) Centrally-planned economy countries	32	59	80	82	82	-
Total	166	281	284	262	249	-
7. Nickel (Thousands of MT)						
a) Latin America	2	11	16	13	11	16
b) Other developing countries	4	10	19	22	23	24
c) Developed countries	316	527	490	435	403	402
d) Centrally-planned economy countries	110	156	188	190	196	-
Total	432	704	713	660	633	-
8. Lead (Thousands of MT)						
a) Latin America	139	253	280	246	236	217
b) Other developing countries	67	152	237	255	302	285
c) Developed countries	2 238	3 338	3 283	3 198	3 141	3 165
d) Centrally-planned economy countries	739	1 281	1 588	1 575	1 588	...
Total	3 183	5 024	5 388	5 274	5 267	...
9. Zinc (Thousands of MT)						
a) Latin America	112	238	348	305	290	299
b) Other developing countries	133	276	435	516	551	519
c) Developed countries	3 070	3 920	3 591	3 470	3 276	3 578
d) Centrally-planned economy countries	782	1 564	1 792	1 786	1 843	-
Total	4 096	5 998	6 166	6 077	5 960	-

	Annual growth rates			Relative breakdown	
	1965-1974	1974-1980	1980-1983	1974	1982

1. Bauxite (Thousands of MT)

a) Latin America	...	7.2	-10.9	2.9	3.4
b) Other developing countries	...	10.4	2.3	3.1	5.9
c) Developed countries	...	0.4	-4.4	74.1	65.9
d) Centrally-planned economy countries	...	3.9	-4.3a/	19.9	24.8
Total	...	1.8	-8.5a/	100.0	100.0

2. Cadmium (MT)

a) Latin America	22.5	13.3	7.6	1.6	3.5
b) Other developing countries	9.8	16.0	9.7	1.1	3.7
c) Developed countries	3.0	-1.6	-0.8	78.4	70.4
d) Centrally-planned economy countries	5.4	3.2	-2.2a/	18.8	22.4
Total	3.6	-	-1.5a/	100.0	100.0

3. Copper (Thousands of MT)

a) Latin America	7.3	7.9	-13.3	3.8	5.0
b) Other developing countries	7.3	12.5	11.4	2.0	4.3
c) Developed countries	2.6	0.4	-1.8	72.1	63.6
d) Centrally-planned economy countries	5.2	4.6	0.5a/	22.1	27.0
Total	3.4	2.1	1.9a/	100.0	100.0

4. Tin (Thousands of MT)

a) Latin America	4.6	3.4	-3.1	3.7	5.0
b) Other developing countries	-	2.8	2.5	4.5	6.0
c) Developed countries	1.3	-3.1	-2.9	70.5	61.5
d) Centrally-planned economy countries	3.0	1.2	-0.9a/	21.3	27.5
Total	1.7	-1.6	-5.1a/	100.0	100.0

5. Iron ore (Millions of MT)

a) Latin America	7.4	-2.8	19.9a/	3.8	5.1
b) Other developing countries	1.6	10.3	8.9a/	3.0	7.1
c) Developed countries	5.1	-2.0	-12.2a/	60.7	45.7
d) Centrally-planned economy countries	4.2	3.5	-2.0a/	32.5	42.1
Total	4.7	0.4	-5.8a/	100.0	100.0

Conclusion Table 8)

	Annual growth rates			Relative breakdown	
	1965-1974	1974-1980	1980-1983	1974	1982

Magnesium (Thousands of MT)

) Latin America	25.1	4.9	-29.3a/	5.3	4.0
) Other developing countries	-	26.0	11.8a/	0.4	2.0
) Developed countries	5.1	-2.2	-8.1a/	73.3	61.0
) Centrally-planned economy countries	7.0	5.2	1.2a/	21.0	32.9
Total	6.0	0.2	-6.4a/	100.0	100.0

Nickel (Thousands of MT)

) Latin America	20.9	6.4	0.0	1.6	1.7
) Other developing countries	10.7	11.3	8.1	1.4	3.6
) Developed countries	5.8	-1.2	-6.4	74.9	63.7
) Centrally-planned economy countries	4.0	3.2	2.1a/	22.1	31.0
Total	5.6	0.2	-5.8a/	100.0	100.0

Lead (Thousands of MT)

) Latin America	6.9	1.7	-8.1	5.0	4.5
) Other developing countries	9.5	7.7	6.3	3.0	5.7
) Developed countries	4.5	-0.3	-1.2	66.4	59.6
) Centrally-planned economy countries	6.3	3.6	0.0a/	25.5	30.1
Total	5.2	1.2	-1.1a/	100.0	100.0

Zinc (Thousands of MT)

) Latin America	8.7	6.5	-4.9	4.0	4.9
) Other developing countries	8.5	7.9	6.1	4.6	9.2
) Developed countries	2.8	-1.5	-0.1	65.3	55.0
) Centrally-planned economy countries	8.0	2.3	1.4a/	26.1	30.9
Total	4.3	0.5	-1.7a/	100.0	100.0

ource: World Bureau of Metal Statistics, "World Metal Statistics Yearbook 1984". UNCTAD "Consideration of International Measures on Iron Ore, Statistical Annex", TD/B/IPC/Iron Ore/2/Add.1, TD/B/IPC/Iron Ore/15.

/ Rate 1980-1982.

Table 9

ESTIMATE OF MINERALS CONSUMPTION - 1980
(In units of metric tons (MT))

Units of consumption Total	Per capita	Minerals	World Total	Per capita	Latin America Total	Per capita	Other Developed Countries Total	Per capita
MT	Gr.	Antimony	64 635	14.86	5 319	15.07	5 181	2.9
Thousands MT	Kg.	Asbestos	4 902	1.13	328	0.93	205	0.1
Thousands MT	Kg.	Barite	7 578	1.74	1 251	3.54	1 397	0.8
Thousands MT	Kg.	Bauxite	93 326	21.46	3 637	10.30	4 774	2.7
MT	Gr.	Bismuth	3 421	0.79	628	1.78	743	0.4
MT	Gr.	Cadmium	16 982	3.90	576	1.63	462	0.2
Thousands MT	Kg.	Chromium	9 729	2.24	418	1.18	1 191	0.6
MT	Gr.	Cobalt	32 724	7.52	493	1.40	660	0.3
Thousands MT	Kg.	Copper	9 422	2.17	497	1.41	337	0.1
Thousands MT	Kg.	Fluorite	4 682	1.08	927	2.63	556	0.3
MT	Gr.	Gold	1 208	0.28	26	0.07	18	0.0
Thousands MT	Kg.	Iron Ore	508 976	117.03	16 000	45.33	27 444	15.8
Thousands MT	Kg.	Lead	5 388	1.24	280	0.79	237	0.1
MT	Gr.	Lithium	6 615	1.52	79	0.22	186	0.1
Thousands MT	Kg.	Magnesium	284	0.07	20	0.06	4	
Thousands MT	Kg.	Manganese	26 697	6.14	2 982	8.45	3 850	2.2
MT	Gr.	Mercury	6 622	1.52	351	0.99	225	0.1
Thousands MT	Kg.	Molybdenum	108	0.02	3	0.01	-	
Thousands MT	Kg.	Nickel	713	0.16	16	0.05	19	0.0
Thousands MT	Kg.	Phosphated rocks	136 014	31.27	6 333	17.94	8 239	4.7
MT	Gr.	Platinum	282	0.06	50	0.14	30	0.0
Thousands MT	Kg.	Potassium	28 855	6.63	1 760	4.99	1 000	0.5
MT	Gr.	Selenium	1 905	0.44	29	0.08	41	0.0
MT	Gr.	Tellurium	508	0.12	1	-	5	
Thousands MT	Kg.	Tin	222	0.05	11	0.03	13	0.0
MT	Gr.	Tungsten	53 320	12.26	1 832	5.19	2 530	1.4
MT	Gr.	Uranium	43 965	10.11	352	1.00	506	0.2
MT	Gr.	Vanadium	33 960	7.81	581	1.65	608	0.3
Thousands MT	Kg.	Zinc	6 166	1.42	348	0.99	435	0.2

Units of consumption		Minerals	World		Developed Countries		Centrally-planned economy countries	
tal	Per capita		Total	Per capita	Total	Per capita	Total	Per capita
	Gr.	Antimony	64 635	14.86	32 979	42.55	21.156	14.22
usands MT	Kg.	Asbestos	4 902	1.13	1 968	2.54	2 401	1.61
usands MT	Kg.	Barite	7 578	1.74	3 366	4.34	1 564	1.05
usands MT	Kg.	Bauxite	93 326	21.46	63 782	82.30	21 133	14.20
	Gr.	Bismuth	3 421	0.79	1 637	2.11	413	0.28
	Gr.	Cadmium	16 982	3.90	12 083	15.59	3 861	2.59
usands MT	Kg.	Chromium	9 729	2.24	4 508	5.82	3 612	2.43
	Gr.	Cobalt	32 724	7.52	26 187	33.79	5 384	3.62
usands MT	Kg.	Copper	9 422	2.17	6 165	7.95	2 423	1.63
usands MT	Kg.	Fluorite	4 682	1.08	1 574	2.03	1 625	1.09
	Gr.	Gold	1 208	0.28	887	1.14	277	0.19
usands MT	Kg.	Iron Ore	508 976	117.03	267 401	345.03	198 131	133.15
usands MT	Kg.	Lead	5 388	1.24	3 283	4.24	1 588	1.07
	Gr.	Lithium	6 615	1.52	4 800	6.19	1 550	1.04
usands MT	Kg.	Magnesium	284	0.07	180	0.23	80	0.05
usands MT	Kg.	Manganese	26 697	6.14	7 779	10.04	12 086	8.12
	Gr.	Mercury	6 622	1.52	3 450	4.45	2 596	1.74
usands MT	Kg.	Molybdenum	108	0.02	92	0.12	13	0.01
usands MT	Kg.	Nickel	713	0.16	490	0.63	188	0.13
usands MT	Kg.	Phosphated rocks	136 014	31.27	88 408	114 07	33 034	22.20
	Gr.	Platinum	282	0.06	100	0.13	102	0.07
sands MT	Kg.	Potassium	28 855	6.63	14 970	19.32	11 125	7.48
	Gr.	Selenium	1 905	0.44	1 578	2.04	257	0.17
	Gr.	Tellurium	508	0.12	423	0.55	79	0.05
sands MT	Kg.	Tin	222	0.05	142	0.18	56	0.04
	Gr.	Tungsten	53 320	12.26	22 436	28.95	26 522	17.82
	Gr.	Uranium	43 965	10.11	43 107	55.62	...	...
	Gr.	Vanadium	33 960	7.81	18 853	24.33	13 918	9.35
sands MT	Kg.	Zinc	6 166	1.42	3 591	4.63	1 792	1.20

ce: 1. See Tables 3 and 8 of the Statistical Appendix.

2. Portney, Paul R. ed. "Current Issues in Natural Resources Policy", Johns Hopkins University Press, Baltimore, 1982.

127

Table 10

LATIN AMERICA: ESTIMATE OF MINERALS CONSUMPTION - 1980
(In units of metric tons (MT))

Minerals	Units of consumption	Total Latin America	Argentina	Brazil	Chile	Colombia	Cuba	Jamaica	Mexico	Peru	Venezuela	Other countries
Antimony	MT	5 319	690	2 220		30	14		2 319	30	30	
Asbestos	Thousands MT	328	18	179	13	20		2	60	8	10	4
Barite	Thousands MT	1 251	65	114	206				334	71	54	407
Bauxite	Thousands MT	3 637	346	1 780		3			673		578	260
Bismuth	MT	628	23	26					576			
Cadmium	MT	576		176					400			
Chromium	Thousands MT	418	5	351		5			51	4	7	
Cobalt	MT	493	138	281					54	19		5
Copper	Thousands MT	497	53	245	43				123	2	1	14
Fluorite	Thousands MT	927	2		4				918	4	2	1
Gold	MT	26	1	11	7						1	
Iron Ore	Thousands MT	16 000	1 356	9 271	1 665				3 708	26		29
Lead	Thousands MT	280	46	83					96			
Lithium	MT	79	26	53								
Magnesium	Thousands MT	20		6					14			
Manganese	Thousands MT	2 982	156	2 203	27	3			492	2	2	97
Mercury	MT	351	60	187		30			68		6	
Molybdenum	MT	3		2					1			
Nickel	Thousands MT	16		11					3			2
Phosphated rocks	Thousands MT	6 333	50	4 060	199	89	152	10	1 621	14	19	119
Platinum	MT	50	8							41		
Potassium	Thousands MT	1 760	39	1 058	141	101	127	16	142	7	5	124
Selenium	MT	29	14			2			13			
Silver	MT	300	45	120		117					15	
Tellurium	MT	1	1									
Tin	Thousands MT	11		5					2			3
Tungsten	MT	1 832	60	1 226	4				291	250	1	
Uranium	MT	352	187	162								3
Vanadium	MT	581		557		24						
Zinc	Thousands MT	348	31	138					89	23	26	41

Source: See Table 8 of the Statistical Appendix.

EVOLUTION OF INTERNATIONAL TRADE, TOTAL
(Billions of dollars FOB)

Exporters \ Importers / Years	Latin America	Canada	European Economic Community	Centrally-planned economy countries	United States	Japan	Other developed countries	Other developing countries	Totals
Africa									
1970	-	-	7	1	1	-	2	1	12
1980	6	-	39	3	30	2	8	6	94
1983	5	-	31	4	14	1	5	9	69
Latin America									
1970	3	1	5	1	6	1	1	-	18
1980	23	3	20	8	37	4	9	6	110
1983	22	3	18	11	33	5	8	8	108
Asia									
1970	-	-	2	1	3	2	2	4	14
1980	4	2	21	6	30	28	8	42	141
1983	4	2	19	6	34	29	10	47	151
Centrally planned-economy countries									
1970	1	-	4	20	-	1	3	4	33
1980	5	-	30	85	2	6	20	27	175
1983	7	-	30	89	3	6	20	35	190
Developed market economies									
1970	14	11	87	9	29	9	41	27	224
1980	76	42	509	61	122	78	156	217	1 261
1983	65	40	433	53	134	70	141	225	1 161
Middle East									
1970	-	-	4	-	-	2	3	2	11
1980	12	3	66	4	20	43	23	40	211
1983	10	-	41	3	9	35	15	44	157
Not classified									
1970	-	-	1	-	-	-	-	-	-
1980	-	-	1	-	-	1	-	-	2
1983	-	-	1	-	-	-	-	-	1
Totals									
1970	18	12	109	32	39	12	52	38	312
1980	126	50	686	167	241	162	224	338	1 994
1983	113	45	573	166	227	146	199	368	1 837

Source: United Nations, Monthly Bulletin of Statistics, May 1984, vol.XXXVIII, N°5.

Table 12

EVOLUTION OF INTERNATIONAL TRADE IN MINERALS AND METALS (SITC revis. 27, 28, 67, 68, 13)
(Millions of dollars FOB)

Exporters	Years	Latin America	Canada	European Economic Community	Centrally-planned economy countries	United States	Japan	Other developed countries	Other developing countries	Totals
Africa	1970	15	2	1 442	105	33	310	346	61	2 314
	1980	148	45	3 064	280	571	555	612	241	5 516
	1983	146	40	1 651	230	612	341	507	439	3 966
Latin America	1970	209	58	1 223	165	992	373	211	47	3 278
	1980	1 450	211	3 813	650	2 148	1 846	852	745	11 715
	1983	814	226	2 853	707	2 168	1 902	580	978	10 228
Asia	1970	14	15	174	90	270	621	72	375	1 631
	1980	114	57	1 688	488	1 716	2 624	510	3 276	10 473
	1983	108	69	1 128	488	1 586	2 312	397	3 060	9 148
Centrally planned-economy countries	1970	104	12	758	2 857	60	186	517	267	4 761
	1980	381	14	2 619	8 725	301	563	1 878	1 336	15 817
	1983	546	13	2 380	9 179	195	581	1 795	2 222	16 911
Developed market economies	1970	1 579	1 056	14 515	1 703	4 557	1 633	5 174	3 451	33 668
	1980	7 555	3 785	63 830	10 583	15 465	6 044	26 547	26 208	160 017
	1983	5 813	3 121	43 249	9 311	13 994	4 237	18 731	26 277	124 733
Middle East	1970	1	-	49	19	10	6	19	9	113
	1980	4	-	168	88	19	34	65	143	521
	1983	2	-	75	147	10	26	52	176	488
Not classified	1970	37	3	110	13	25	99	29	32	348
	1980	277	3	772	26	129	415	639	228	2 489
	1983	267	4	372	16	165	475	833	1 440	3 572
Totals	1970	1 959	1 146	18 271	4 952	5 947	3 228	6 368	4 242	46 113
	1980	9 929	4 115	75 954	20 840	20 349	12 081	31 103	32 177	206 548
	1983	7 696	3 473	51 708	20 078	18 730	9 874	22 895	34 592	169 046

EVOLUTION OF INTERNATIONAL TRADE IN MINERAL CONCENTRATES AND SCRAP (SITC revis. 27, 28)
(Millions of dollars FOB)

Exporters	Years	Latin America	Canada	European Economic Community	Centrally-planned economy countries	United States	Japan	Other developed countries	Other developing countries	Totals
Africa	1970	5	1	426	47	28	61	67	11	646
	1980	130	45	1 824	240	183	212	451	163	3 248
	1983	102	30	1 086	163	242	121	366	170	2 280
Latin America	1970	15	58	359	162	597	255	113	30	1 589
	1980	145	110	1 604	414	1 138	1 085	591	474	5 561
	1983	112	194	1 476	539	983	1 177	356	431	5 268
Asia	1970	1	1	52	42	36	472	43	134	781
	1980	1	3	341	182	162	1 697	171	510	3 067
	1983	2	4	146	156	100	1 301	156	448	2 313
Centrally planned economy countries	1970	13	1	171	733	13	73	107	14	1 125
	1980	44	1	572	2 735	84	205	469	118	4 228
	1983	91	-	509	1 955	62	196	503	209	3 525
Developed market economies	1970	76	155	2 856	93	756	1 205	708	235	6 084
	1980	350	901	10 520	701	2 045	3 621	3 567	2 180	23 885
	1983	341	607	6 746	506	1 549	2 592	2 496	1 841	16 678
Middle East	1970	1	-	49	19	10	6	19	9	113
	1980	4	-	168	88	19	34	65	143	521
	1983	2	-	75	147	10	26	52	176	488
Not classified	1970	-	3	19	-	1	94	2	-	119
	1980	1	-	141	-	-	372	56	-	569
	1983	1	-	131	-	1	249	24	14	420
Totals	1970	111	219	3 932	1 096	1 441	2 166	1 059	433	10 457
	1980	674	1 060	15 170	4 360	3 631	7 226	5 370	3 588	41 079
	1983	651	835	10 169	3 466	2 947	5 662	3 953	3 289	30 972

Source: See Table 11 of the Statistical Appendix.

Table 14

EVOLUTION OF THE VOLUME OF EXPORTS OF THE MAIN MINERALS AND METALS
(Annual growth rates)

Main Exporters / Importers	Periods	Total	Developed countries	Developing countries	Africa	Latin America	Asia	Other developing countries	Centrally planned economy countries
I. Aluminium									
Europe	1970-77	1.3	2.7	-10.0	-6.6	-8.8	34.3	-28.1	2.8
	1977-81	0.2	1.6	-3.2	-4.7	7.8	-61.2	-14.7	-23.8
Japan	1970-77	8.4	4.3	28.5	14.8	...	41.7	14.9	15.5
	1977-81	23.0	20.9	36.8	13.1	70.9	4.2	...	-9.1
United States	1970-77	9.8	6.4	127.6	...	98.2	...	-4.4	...
	1977-81	1.4	2.1	-2.1	-3.2	21.8	-61.5	-10.2	...
Australia	1970-77	-10.8	-7.9	-28.1	-	-	-	-28.1	-
	1977-81	166.7	147.2	271.2	-	-	...	-43.8	...
Total	1970-77	3.8	3.7	3.9	-6.2	-2.1	44.5	-27.6	5.5
	1977-81	5.1	4.8	13.5	-0.8	60.8	-1.9	-14.9	-17.4
II. Bauxite									
Europe	1970-77	9.6	4.8	18.5	27.1	1.0	-32.5	-23.8	-18.3
	1977-81	0.9	-5.3	5.5	5.0	10.1	12.7	65.9	55.0
Soviet Union	1970-77	13.6	-2.5	55.2	55.2				
	1977-81	-7.6	-14.4	-4.9	-9.7				
Japan	1970-77	5.5	9.4	0.1		9.9	...	...	38.9
	1977-81	-4.9	-7.5	-0.8		-10.6	-0.2	-36.7	22.5
United States	1970-77	-1.2	-0.8	-1.2	103.9	-3.7	-0.4	...	
	1977-81	1.1	-24.9	0.8	13.0	-2.2	...	-2.4	...
Canada	1970-77	1.2	5.4	1.2	66.5	-9.5	-34.7	...	
	1977-81	-0.5	6.7	-0.6	-19.5	15.4	...	...	
Total	1970-77	3.3	5.0	2.8	41.4	-4.2	-3.4	-18.4	-13.1
	1977-81	-0.8	-7.2	0.9	1.2	0.6	0.8	49.4	69.3

Main Exporters \ Importers \ Periods	Total	Developed countries	Developing countries	Africa	Latin America	Asia	Other developing countries	Other Centrally-planned economy countries
III. Metallic copper (smelted and refined)								
Europe 1970-77	1.7	2.3	0.2	0.03	0.9	-46.4	0.2	28.1
1977-81	-3.7	-4.0	-3.9	-7.4	1.1	79.1	-25.1	0.4
Soviet Union 1970-77	50.6	108.6	-	-	-	-	-	...
1977-81	-4.2	-	-	-	-	-	-	-29.7
Japan 1970-77	-1.4	5.7	-2.3	-4.7	4.5	-10.4	-	...
1977-81	4.9	7.9	4.3	0.8	9.7	38.5	-	...
United States 1970-77	2.9	3.3	2.7	70.6	-2.2	-	19.1	-
1977-81	2.1	-8.3	7.3	-0.5	10.1	-	26.3	-
Total 1970-77	1.6	2.5	0.1	0.1	0.5	-32.8	-2.5	30.7
1977-81	-2.1	-3.8	-1.0	-5.5	4.3	48.7	-23.1	-2.0
IV. Copper ore (concentrates)								
Europe 1970-77	13.2	8.6	16.6	11.0	-1.1	11.5	114.1	
1977-81	-2.9	-13.4	1.7	-21.4	21.6	-23.7	-1.4	-
Japan 1970-77	8.8	5.5	11.0	10.1	5.3	7.9	166.5	-
1977-81	4.3	8.8	1.5	-5.7	13.5	1.0	-4.5	-
United States 1970-77	8.2	23.4	3.2	-	-6.3	-0.5	-39.2	-
1977-81	-11.2	-38.4	-1.6	-	28.5	-1.5	-	-
Total 1970-77	9.6	6.4	11.8	12.2	2.5	7.8	123.9	-
1977-81	2.5	4.2	1.5	-10.7	16.7	-0.4	-3.1	-

(Cont. Table 14)

Periods	Main Exporters / Importers	Total	Developed countries	Developing countries	Africa	Latin America	Asia	Other developing countries	Centrally-planned economy countries
V. Metallic tin									
1970-77	Europe	-0.3	-6.5	2.2	-9.2	...	3.8	10.6	13.2
1977-81		-1.2	-4.3	-0.4	-14.0	5.6	-1.3	26.5	-11.4
1970-77	Japan	0.7	...	0.9	-	...	0.9	-44.2	-13.8
1977-81		1.7	-	1.6	-	...	1.6	177.1	34.4
1970-77	Soviet Union	2.2	-8.0	11.8	-	42.7	-2.9	25.3	...
1977-81		10.4	-17.8	16.6	-	-39.8	51.9	-4.1	...
1970-77	United States	-1.0	-4.2	-1.1	-54.4	46.5	-2.7	-0.1	...
1977-81		-1.0	-3.8	-1.9	301.5	13.2	-5.2	38.2	51.9
1970-77	Canada	-0.2	20.5	-17.2	...	...	-44.1	103.2	-
1977-81		-6.7	-15.0	12.9	-	8.0	-17.6	103.2	-
1970-77	Total	-0.2	-3.4	0.5	-10.3	44.3	-0.5	13.3	-10.8
1977-81		0.2	-9.2	0.9	-9.3	2.4	0.3	19.8	17.2
VI. Tin concentrates									
1970-77	Europe	-5.4	9.4	-7.1	-3.0	-9.2	29.9	-18.1	-
1977-81		-15.8	-16.9	-15.6	-6.4	-24.2	-45.1	-19.4	-
1970-77	United States	5.1	-	5.1	-	5.0	-	68.9	-
1977-81		-48.7	-	-48.7	-	...	-	...	-
1970-77	Total	-3.7	9.4	-4.9	-3.0	-5.5	29.9	-17.6	-
1977-81		-20.3	-16.9	-20.9	-6.4	-23.9	-45.1	-6.5	-

(Cont. Table 14)

Main Importers \ Exporters \ Periods	Total	Developed countries	Developing countries	Africa	Latin America	Asia	Other developing countries	Centrally-planned economy countries
VII. Metallic lead (pig lead)								
1970-77 Europe	-0.6	0.2	-3.2	-7.9	4.8	3.8	3.4	-9.1
1977-81	-1.9	-0.4	-9.3	3.2	-19.6	-43.3	-29.2	-42.6
1970-77 Soviet Union	-2.3	2.1	65.7	..	-22.9	..	..	..
1977-81	-9.0	-3.1	-22.9	-	...	..	..	..
1970-77 Japan	49.7	54.7	52.0	44.0	42.1	21.9	...	46.7
1977-81	19.2	29.6	35.9	34.6	3.1	-13.3	-	-2.1
1970-77 United States	0.6	-1.1	3.1	-	3.1	3.8	-	-
1977-81	-18.2	-15.3	-22.4	-	-22.4	...	-22.8	...
1970-77 Total	0.3	0.2	2.2	-7.3	4.6	10.3	0.8	-6.1
1977-81	-4.7	-2.3	-12.4	5.5	-16.2	-24.8	-28.8	-17.7
VIII. Lead ore (concentrates)								
1970-77 Europe	-6.8	-5.7	-9.3	-2.0	-15.0	-16.4	-57.8	-11.6
1977-81	3.9	2.2	9.3	-	13.0	47.0	42.7	-6.7
1970-77 Japan	-1.1	-1.4	-0.3	..	3.6	-6.7	-	..
1977-81	7.2	5.7	10.9	-	10.0	13.4	-	-
1970-77 United States	-5.9	-9.2	-1.6	-	-1.6	...	144.5	-
1977-81	-16.0	-12.3	-35.3	-	-35.8	...	-57.8	-
1970-77 Total	-5.4	-5.1	-5.6	-2.4	-5.8	-10.6	-57.8	-11.4
1977-81	2.5	1.9	2.6	-	-1.4	26.7	151.3	25.4

(Cont. Table 14)

Periods	Main Importers \ Exporters	Total	Developed countries	Developing countries	Africa	Latin America	Asia	Other developing countries	Centrally-planned economy countries
IX. Metallic zinc									
1970-77	Europe	8.7	9.7	-6.4	0.9	16.5	...	13.1	4.4
1977-81		-3.4	1.1	-31.4	-34.3	-27.6	...	-18.5	-41.0
1970-77	Soviet Union	-4.7	...	...	-	-	-	-	-17.4
1977-81		-2.1	1.4	...	-	-	-	-	-9.9
1970-77	Japan	4.0	-36.3	-6.8	...	-5.4	-22.9	...	20.7
1977-81		1.9	7.5	60.5	-	-22.8	236.3	-	-4.9
1970-77	United States	11.4	12.5	8.3	20.5	3.2	-	-0.8	-5.9
1977-81		3.6	3.6	4.4	-2.8	8.3	-	8.8	-17.8
1970-77	Total	8.9	11.0	6.9	9.1	5.8	21.8	-17.2	-0.4
1977-81		0.1	2.3	-2.8	-11.8	0.8	58.0	-10.7	-22.0
X. Zinc ore (concentrates)									
1970-77	Europe	2.2	-	10.1	-6.1	18.2	5.8	-56.4	1.7
1977-81		-0.9	0.2	-3.7	-13.2	-1.4	-21.4	156.5	-10.0
1970-77	Japan	-0.1	2.7	-1.9	...	-1.6	-1.6	...	-12.0
1977-81		-2.3	0.9	-8.7	-	-3.9	-42.3	-	18.6
1970-77	United States	-18.8	-20.8	-16.0	-	-19.6	...	-25.0	-
1977-81		1.4	2.5	0.2	-	8.2	...	-25.2	-
1970-77	Total	-0.1	-0.9	2.0	-7.3	3.6	2.4	-55.6	-9.3
1977-81		-1.2	0.4	-5.2	-13.2	-1.8	-32.4	147.2	12.1

(Conclusion Table 14)

Periods	Importers	Main Exporters Total	Developed countries	Developing countries	Africa	Latin America	Asia	Not classified
XI. Iron ore								
1965-75	Europe	2.7	8.3	4.2	0.2	9.5	-6.3	-0.8
1975-78		-0.6	0.1	-0.1	0.1	0.4	-22.2	-11.1
1978-81		-0.3	-3.0	4.2	2.4	5.9	-81.1	-21.3
1965-75	United States	0.4	-1.6	2.0	-1.7	2.5	. . .	15.0
1975-78		-10.4	-0.2	-19.5	-4.6	-21.3	. . .	-67.0
1978-81		-5.6	-2.1	-11.3	-0.2	-13.8	. - .	-30.7
1965-75	Japan	12.9	32.8	9.8	19.2	10.6	7.8	-0.2
1975-78		-4.5	-2.1	-5.4	-25.1	-4.8	-5.1	-13.5
1978-81		2.5	2.2	5.3	15.9	6.1	2.9	-11.4
1965-75	Total	5.4	6.4	5.5	0.5	7.4	5.9	0.4
1975-78		3.8	-0.9	-5.5	-1.5	-6.5	-6.4	-13.5
1978-81		0.3	-0.6	3.0	2.6	3.4	1.5	-14.9
XII. Raw steel								
1973-78	Europe	1.2	1.2	-	-	-	-	-
1978-82		-0.9	-0.9	-	-	-	-	-
1973-78	United States	7.0	6.9	7.8	-	7.8	-	-
1978-82		-5.2	-5.6	1.2	-	1.2	-	-
1973-78	Other countries	8.4	7.6	17.1	-	10.9	20.1	-
1978-82		-1.3	-4.6	21.3	-	21.9	21.2	-
1973-78	Total	4.3	3.9	15.0	-	9.6	20.1	-
1978-82		-1.3	-2.7	20.1	-	14.9	21.2	-

Source: - Metallgesellschaft Aktiengesellschaft, "Metal Statistics", several issues.
- Association of Iron Ore Exporting Countries, "Iron Ore Statistics", September 1983.
- International Iron and Steel Institute Committee on Statistics, "Steel Statistics Yearbook 1983".
- (ISSN 0771-2871, Brussels 1984).

Table 15

LATIN AMERICA: SELECTED METALS EXPORTS BY MAIN EXPORTING COUNTRIES

(In thousands of current dollars FOB)

	1970	1971	1972	1973	1974	1975	1976
Bauxite/Alumina/Aluminium	385 926	378 794	373 034	402 250	731 887	782 693	788 012
Brazil	129	198	255	149	671	695	194
Guyana	69 279	68 253	63 351	64 317	90 007	112 425	113 788
Jamaica	224 279	217 778	234 985	249 959	504 929	499 087	427 536
Mexico	393	1 563	683	231	111	63	
Dominican Republic	15 132	15 983	14 864	14 835	17 756	16 725	15 521
Venezuela	7 414	6 713	4 950	5 837	14 043	10 098	11 373
Suriname	69 300	68 306	53 946	66 922	104 370	143 600	219 600
Copper	1 271 081	884 098	867 866	1 335 300	2 212 968	1 116 253	1 495 613
Bolivia	12 499	8 297	8 770	13 440	16 018	7 263	6 519
Brazil	140	167	1 057	1 159	310	102	
Chile	973 167	685 524	628 941	990 584	1 806 965	908 613	1 241 797
Mexico	8 225	11 200	32 480	38 264	20 992	25 296	13 750
Peru	277 050	178 910	196 618	291 853	368 683	174 979	233 547
Tin	106 367	109 944	119 106	136 967	253 039	197 836	231 663
Bolivia	102 047	105 878	113 541	130 993	230 117	171 398	216 329
Brazil	4 051	3 750	5 262	5 755	21 440	24 123	13 716
Mexico						769	
Peru	269	316	303	219	1 482	1 546	1 618
Iron Ore	534 354	537 786	506 077	692 317	1 082 606	1 413 782	1 557 639
Brazil	218 767	243 237	243 478	386 384	602 102	986 609	1 073 973
Chile	71 983	69 745	56 012	56 629	132 444	88 885	106 301
Mexico					41	353	413
Peru	66 455	62 153	64 950	60 710	60 099	55 053	56 011
Venezuela	177 149	162 651	141 637	188 594	287 920	282 882	320 941

(Cont. Table 15)

	1970	1971	1972	1973	1974	1975	1976
Nickel	104 700	107 516	160 433	201 847	223 355	272 686	294 268
Brazil		516			558		
Chile							
Dominican Republic			47 133	83 447	93 097	102 186	110 768
Cuba	104 700	107 000	113 300	118 400	129 700	170 500	183 500
Silver	73 904	34 457	61 588	226 234	233 008	277 217	241 342
Bolivia	10 531	8 342	7 590	12 561	26 834	28 541	24 323
Brazil	124	140	270	672	1 476	506	787
Chile	5 317	4 759	2 357	815	10 047	33 791	8 756
Mexico	29 187	215	19 937	186 324	112 331	131 381	114 566
Peru	28 745	21 001	31 434	25 862	82 320	82 998	92 910
Lead	99 095	74 798	81 512	110 566	201 237	127 649	142 689
Bolivia	7 806	5 948	5 776	8 347	11 495	7 706	8 436
Brazil	513						
Chile	172		224			353	274
Mexico	27 132	19 374	20 717	23 894	71 333	46 101	40 082
Peru	63 472	49 476	54 795	78 325	118 409	73 489	93 897
Zinc	96 599	94 376	125 830	152 536	351 431	308 623	329 467
Bolivia	14 319	15 270	15 438	25 963	37 657	40 332	39 139
Brazil					1 321	1 675	1 339
Chile				326	1 437	1 286	2 702
Mexico	35 038	31 380	37 432	28 165	136 003	92 666	114 524
Peru	47 242	47 726	72 960	98 082	175 013	172 664	171 763
Total	2 672 026	2 221 769	2 295 446	3 258 017	5 289 531	4 496 739	5 080 693

	1977	1978	1979	1980	1981	1982
Bauxite/Alumina/Aluminium	960 503	962 534	1 201 330	1 748 128	1 806 228	1 511 375
Brazil	177	202	11 920	65 180	116 980	92 325
Guyana	129 905	98 148	129 469	170 431	152 200	94 500
Jamaica	538 092	539 938	581 673	735 700	760 200	544 200
Mexico						
Dominican Republic	22 017	23 143	20 902	18 513	15 648	5 250
Venezuela	7 612	18 703	143 666	335 304	383 900	445 300
Suriname	262 700	282 400	313 700	423 000	377 500	329 800
Copper	1 465 180	1 563 305	2 545 845	2 985 094	2 571 195	2 410 956
Bolivia	4 099	3 968	3 296	3 536	4 402	3 126
Brazil		153	833	1 162	16 383	315
Chile	1 058 501	1 161 910	1 799 600	2 200 400	1 714 900	1 731 400
Mexico	31 482	6 719	53 881	155 722	306 810	218 615
Peru	371 098	390 555	688 235	624 274	528 700	457 500
Tin	351 064	403 533	432 533	432 405	415 512	350 063
Bolivia	326 653	373 678	396 567	378 149	343 095	278 343
Brazil	21 130	21 058	23 293	46 547	64 517	55 920
Mexico						:::
Peru	3 281	8 797	13 673	7 709	7 900	15 800
Iron Ore	1 382 866	1 333 724	1 898 289	2 433 665	2 668 380	2 495 522
Brazil	945 055	900 914	1 419 686	1 864 481	2 060 780	2 027 322
Chile	106 601	118 997	110 400	157 600	161 900	158 200
Mexico	8 749	12 936	18 222	18 000	18 000	18 000
Peru	84 640	69 191	85 916	80 558	93 300	105 200
Venezuela	237 821	231 686	264 065	313 026	334 400	186 800

(Conclusion Table 15)

	1977	1978	1979	1980	1981	1982
Nickel						
Brazil	283 421	233 895	316 515	389 527	415 244	289 906
Chile			118	...	...	396
Dominican Republic	91 921	73 495	123 097	101 927	110 544	24 210
Cuba	191 500	160 400	193 300	287 600	304 700	265 300
Silver						
Bolivia	269 351	296 826	331 213	316 135	466 108	324 108
Brazil	30 808	33 764	58 267	118 328	71 693	37 067
Chile	361	882	1 043	3 162	15	141
Mexico	19 880	31 926	48 600	120 000	82 000	81 500
	103 407	128 181	333			
Peru	114 895	102 073	222 970	74 645	312 400	205 400
Lead						
Bolivia	186 828	213 605	389 264	410 910	237 286	208 487
Brazil	12 398	10 683	17 992	14 450	11 459	6 540
Chile	1 101	258	201	859	135	
Mexico	51 677	64 542	80 800	63 141	33 592	22 047
Peru	121 652	138 122	290 271	332 460	192 100	179 900
Zinc						
Bolivia	278 311	234 137	298 744	335 434	392 939	365 994
Brazil	44 745	31 362	42 678	36 679	40 423	38 295
Chile	925	460	1 352	-	-	2 002
	2 002	594				
Mexico	91 106	100 346	100 777	107 483	85 116	57 979
Peru	139 533	101 375	153 937	191 272	267 400	267 900
Total	5 177 524	5 241 559	7 413 733	9 051 298	8 972 892	7 956 411

Source: U.N. Commodity Trade Statistics, Statistical Papers, various issues; Foreign Trade Yearbooks, Central Bank Bulletins, various country statistical publications.

Table 16

LATIN AMERICA: SELECTED METALS EXPORTS BY MAIN EXPORTING COUNTRIES

(In thousands of 1975 dollars) a/

	1970	1971	1972	1973	1974	1975	1976
Bauxite/Alumina/Aluminium	778 076	708 027	651 020	569 760	833 583	782 693	775 604
Brazil	620	370	445	211	764	695	191
Guyana	139 675	127 576	110 560	91 101	102 514	112 425	111 996
Jamaica	452 175	407 062	410 096	354 050	575 090	499 087	420 803
Mexico	792	2 921	1 192	327	126	63	
Dominican Republic	30 508	29 875	25 941	21 013	20 223	16 725	15 277
Venezuela	14 948	12 548	8 639	8 268	15 994	143 600	11 194
Suriname	139 718	127 675	94 147	94 790	118 872	10 098	216 142
Copper	2 562 664	1 652 519	1 514 600	1 891 360	2 520 465	1 116 253	1 472 059
Bolivia	25 200	15 508	15 305	19 037	18 244	7 263	6 416
Brazil	282	312	1 845	1 642	353	102	
Chile	1 962 030	1 281 353	1 097 628	1 403 093	2 058 047	908 613	1 222 241
Mexico	16 583	20 935	56 684	54 198	23 909	25 296	13 533
Peru	558 569	334 411	343 138	413 390	419 912	174 979	229 869
Tin	214 449	205 503	207 864	194 004	288 199	197 836	228 015
Bolivia	205 740	197 903	198 152	185 542	262 092	171 398	212 922
Brazil	8 167	7 009	9 183	8 152	24 419	24 123	13 500
Mexico	-	-	-	-		769	
Peru	542	591	529	310	1 688	1 546	1 593
Iron Ore	1 077 327	1 005 208	883 206	980 619	1 233 036	1 413 782	1 533 109
Brazil	441 063	454 649	424 918	547 286	685 765	986 609	1 057 060
Chile	145 127	130 364	97 752	80 211	150 847	88 885	104 627
Mexico	-	-	-	-	47	353	406
Peru	133 982	116 174	113 351	85 992	68 450	55 053	55 129
Venezuela	357 155	304 021	247 185	267 130	327 927	282 882	315 887

(Cont. Table 16)

	1970	1971	1972	1973	1974	1975	1976
Nickel	211 089	200 964	279 988	285 902	254 391	272 686	289 634
Brazil	-	-	-	-	-	-	-
Chile	-	-	-	-	636	-	-
Cuba	211 089	200 000	197 731	167 705	147 722	170 500	180 610
Dominican Republic		964	82 257	118 197	106 033	102 186	109 024
Silver	149 001	64 406	107 483	320 455	265 386	277 217	237 542
Bolivia	21 232	15 593	13 246	17 792	30 563	28 541	23 940
Brazil	250	262	471	952	1 681	506	775
Chile	10 720	8 895	4 113	1 154	11 443	33 791	8 618
Mexico	58 845	402	34 794	263 915	127 940	131 381	112 762
Peru	57 954	39 254	54 859	36 632	93 759	82 998	91 447
Lead	199 789	139 810	142 254	156 609	229 199	127 649	140 442
Bolivia	15 738	11 118	10 080	11 823	13 092	7 706	8 303
Brazil	1 034	-	-	-	-		-
Chile	347		391		-	353	270
Mexico	54 702	36 213	36 155	33 844	81 245	46 101	39 451
Peru	127 968	92 479	95 628	110 942	134 862	73 489	92 418
Zinc	194 755	176 403	219 598	216 057	400 264	308 623	324 278
Bolivia	28 868	28 542	26 942	36 775	42 890	40 332	38 523
Brazil					1 505	1 675	1 318
Chile				462	1 637	1 286	2 659
Mexico	70 641	58 654	65 326	39 894	154 901	82 666	112 720
Peru	95 246	89 207	127 330	138 926	199 331	172 664	169 058
Total	5 387 150	4 152 840	4 006 013	4 614 756	6 024 523	4 496 739	5 000 683

(Cont. Table 16)

	1977	1978	1979	1980	1981	1982
Bauxite/Alumina/Aluminium	873 979	739 842	832 523	1 117 016	1 216 315	1 039 461
Brazil	161	155	8 261	41 649	78 774	63 497
Guyana	118 203	75 440	89 722	108 902	102 357	64 993
Jamaica	489 620	415 018	403 100	470 096	511 919	374 278
Mexico		-				-
Dominican Republic	20 034	17 789	14 485	11 829	10 537	3 611
Venezuela	6 926	14 376	99 561	214 252	258 519	306 259
Suriname	239 035	217 064	217 394	270 288	254 209	226 823
Copper	1 333 194	1 201 618	1 764 272	1 907 407	1 731 444	1 658 154
Bolivia	3 730	3 050	2 284	2 259	2 964	2 150
Brazil		118	577	742	11 032	217
Chile	963 149	893 090	1 247 124	1 406 006	1 154 815	1 190 784
Mexico	28 646	5 164	37 340	99 503	206 606	150 354
Peru	337 669	300 196	476 947	398 897	356 027	314 649
Tin	319 439	310 172	299 745	276 297	279 806	240 759
Bolivia	297 227	287 224	274 128	241 629	231 040	191 433
Brazil	19 227	16 186	16 142	29 742	43 446	38 459
Mexico						
Peru	2 985	6 762	9 475	4 926	5 320	10 867
Iron Ore	1 258 295	1 025 153	1 315 515	1 555 059	1 796 889	1 716 315
Brazil	859 923	692 478	983 843	1 191 362	1 387 731	1 394 307
Chile	96 998	91 466	76 507	100 703	109 024	108 803
Mexico	7 961	9 943	12 628	11 502	12 121	12 380
Peru	77 015	53 183	59 540	51 475	62 828	72 352
Venezuela	216 398	178 083	182 997	200 017	225 185	128 473

	1977	1978	1979	1980	1981	1982
Nickel	257 890	179 781	219 345	248 899	279 625	199 385
Brazil	-	.	82	.	.	272
Chile	174 249	123 290	133 957	183 770	205 185	182 462
Cuba	83 641	56 491	85 306	65 129	74 440	16 651
Dominican Republic						
Silver	245 087	228 152	229 531	202 002	313 877	222 907
Bolivia	28 033	25 952	40 379	75 609	48 278	25 493
Brazil	328	678	723	2 020	10	97
Chile	18 089	24 540	33 680	76 677	55 219	56 052
Mexico	94 092	98 525	231			
Peru	104 545	78 457	154 518	47 696	210 370	141 265
Lead	169 998	164 185	269 759	262 563	159 788	143 389
Bolivia	11 281	8 211	12 468	9 233	7 716	4 498
Brazil						
Chile	1 002	198	139	549	91	
Mexico	47 022	49 610	55 994	40 346	22 621	15 163
Peru	110 693	106 166	201 158	212 435	129 360	123 728
Zinc	253 241	179 968	207 030	214 335	264 605	251 715
Bolivia	40 714	24 106	29 576	23 437	27 221	26 338
Brazil	842	354	937			1 377
Chile	1 822	457				
Mexico	82 899	77 130	69 839	68 679	57 317	39 750
Peru	126 964	77 921	106 678	122 219	180 067	184 250
Total	4 711 123	4 028 871	5 137 720	5 783 578	6 042 349	5 472 085

Source: See Table 15 of the Statistical Appendix.

a/ Deflator: Manufacturing Unit Value (MUV) index, FOB (IBRD, Commodity Trade and Price Trends, 1983-1984 Edition).

Table 17

LATIN AMERICA: EVOLUTION OF THE COMPOSITION OF
SELECTED MINERALS AND METAL EXPORTS a/
(In thousands of 1975 dollars)

	1970	1974	1980	1982
Bolivia	296 778	366 881	362 167	249 912
Brazil	451 056	714 487	1 266 064	1 498 226
Chile	2 118 224	2 222 610	1 583 386	1 355 639
Cuba	211 089	147 722	183 770	182 462
Dominican Republic	30 508	126 256	76 958	20 262
Guyana	139 675	102 514	108 902	64 993
Jamaica	452 175	575 090	470 096	374 278
Mexico	201 563	388 168	220 030	217 647
Peru	974 261	918 002	837 648	847 111
Suriname	139 718	118 872	270 288	226 823
Venezuela	372 103	343 921	414 269	434 732
Total	5 387 150	6 024 523	5 783 578	5 472 085
Percentage with an intra-regional destination	4.1	6.4	11.2 b/	

(Conclusion Table 17)

	Percentage structure				Annual growth rates		
	1970	1974	1980	1982	1970-1974	1974-1980	1980-1982
Bolivia	5.5	6.1	6.1	4.6	5.4	-0.7	-15.8
Brazil	8.4	11.9	21.9	27.4	12.2	10.0	8.8
Chile	39.3	36.9	27.4	24.8	1.2	-5.5	-7.5
Cuba	3.9	2.5	3.2	3.3	-8.5	3.7	-0.4
Dominican Republic	0.6	2.1	1.3	0.4	42.6	-7.9	-48.7
Guyana	2.6	1.7	1.9	1.2	-7.4	1.0	-22.7
Jamaica	8.4	9.5	8.1	6.8	6.2	-3.3	-10.8
Mexico	3.7	6.4	3.8	4.0	17.8	-9.0	-0.5
Peru	18.3	15.2	14.5	15.5	-1.5	-1.5	0.6
Suriname	2.6	2.0	4.7	4.1	-4.0	14.7	-8.4
Venezuela	6.9	5.7	7.2	7.9	-1.9	3.2	2.4
Total	100.0	100.0	100.0	100.0	2.8	-0.7	-2.7

Source: See Table 15 of the Statistical Appendix.

a/ Aluminium, copper, iron ore, tin, nickel, silver, lead and zinc.
b/ 1979.

147

Table 18

RELATIVE IMPORTANCE OF MINING IMPORTS IN TOTAL SUPPLY (1970-1982)

(USA, EEC, Japan, USSR)

(Percentages)

Minerals Metals	United States	European Community	Japan	USSR	Main exporting countries	Potential for trade complementarity with Latin America (depending on reserves)
Alumina	-	84	13	-	Australia, Canada, Jamaica, Suriname	-
Aluminium	-	28	31	(Less than 1)	Canada, United States, Australia	-
Antimony	51	91	100	20	South Africa, Bolivia, China, Mexico	Bolivia, Mexico, Peru
Asbestos	80	82	99	(Less than 1)	Canada, South Africa, China, Zimbabwe	Mexico, Colombia, Venezuela
Barite	43	(Less than 1)	43	50	Peru, China, Mexico, Morocco, Chile	Peru, Mexico, Chile
Bauxite	94	-	100	60	Jamaica, Australia, Guinea, Suriname	Brazil, Jamaica, Guyana, Suriname
Cadmium	63	53	(Less than 1)	5	Canada, Australia, Mexico, Belgium, Luxembourg	Mexico
Cobalt	91	100	100	43	Zaire, Japan, Canada, Zambia, Finland	Cuba
Copper	5	99	87	(Less than 1)	Chile, Canada, Zambia, Zaire, Peru	Chile, Peru, Mexico, Panama
Columbium	100	100	100	(Less than 1)	Brazil, Canada, Nigeria, Australia	Brazil
Chromium	90	99	99	(Less than 1)	South Africa, Brazil, USSR, Finland	Brazil, Cuba
Fluorite	85	18	100	47	Mexico, South Africa, Thailand, Kenya	Mexico
Graphite	100	84	-		Mexico, Korea, Madagascar, USSR, India	Mexico
Gold	7	99	94	(Less than 1)	Canada, USSR, South Africa, Brazil	Dominican Republic, Colombia, Mexico
Gypsum	37	(Less than 1)	(Less than 1)	(Less than 1)	Canada, Mexico, Egypt, Australia	Mexico
Industrial diamonds	100	-	-	-	Ireland, South Africa, Belgium, Luxembourg, England	...
Iron Ore	28	79	99	(Less than 1)	Australia, Brazil, Canada, India, South Africa	Bolivia, Brazil, Cuba

Minerals Metals	United States	European Community	Japan	USSR	Main exporting countries	Potential for trade complementarity with Latin America (depending on reserves)
Manganese	98	99	97	(Less than 1)	South Africa, Gabon, India, Brazil	Brazil, Bolivia
Mercury	39	86	(Less than 1)	(Less than 1)	Spain, Algeria, Mexico, Turkey, Italy	Mexico
Mica	100	83	100	2	India, Brazil, Madagascar, Korea	Brazil
Molybdenum	-	100	99	(Less than 1)	Canada, Chile, Peru	Chile, Peru
Nickel	72	100	100	(Less than 1)	Australia, Canada, New Caledonia, Cuba, Indonesia	Cuba, Dominican Republic
Lead	10	70	75	(Less than 1)	Canada, Mexico, Peru, Australia	Mexico, Peru
Phosphated rocks	-	99	100	(Less than 1)	Jordan, Morocco, South Africa, Tunisia	Mexico, Brazil
Platinum	85	100	98	(Less than 1)	South Africa, USSR, Canada, Colombia	Colombia
Potasium	68	1	-	(Less than 1)	Canada, Israel, Spain, Germany	-
Salt	-	(Less than 1)	87	(Less than 1)	India, Mexico, Australia, Brazil	Mexico, Brazil
Selenium	49	100	(Less than 1)	(Less than 1)	Canada, Japan, Yugoslavia	Chile, Peru, Mexico
Silver	50	98	73	2	Peru, Mexico, Canada, Australia	Mexico, Peru
Steel	19	(Less than 1)	-	1	Japan, Europe, Canada, Brazil	-
Strontium	100	30	-	(Less than 1)	Mexico, Spain, Turkey, England	Mexico
Sulphur	7	26	(Less than 1)	(Less than 1)	Canada, Mexico, Spain	Mexico
Tantalum	91	100	100	(Less than 1)	Australia, Canada, Thailand, Brazil	Brazil
Tin	80	95	96	11	Malaysia, Indonesia, Bolivia, Thailand	Bolivia, Brazil
Titanium (ilmenite rutile)	43	100	100	(Less than 1)	Australia, Norway, India, Malaysia	Brazil
Tungsten	52	77	75	2	China, Korea, Canada, Bolivia, Australia	Bolivia, Mexico, Brazil
Vanadium	42	100	100	(Less than 1)	South Africa, China, Finland, Australia	Chile, Venezuela
Zinc	67	71	59	(Less than 1)	Canada, Australia, Peru, Mexico, Spain	Peru, Mexico

Source: See Table 9 of the Statistical Appendix.

Table 19

EVOLUTION OF THE INTERNATIONAL PRICES OF MINERALS
(Dollars per unit of high grade ore)

Unit	Mineral	1947	1960	1965	1974	1975	1978	1980	1981	1982	1983
Pound	Antimony	0.33	0.42	0.46	1.82	1.77	1.14	1.51	1.36	1.07	2.50
Pound	Arsenic (dioxide)	0.06	0.06	0.06	0.07	0.23	0.23	0.32	0.40	0.40	...
Kg	Asbestos (spinning)	0.32	0.64	0.84	0.79	0.90	0.90	0.84	0.93	0.92	...
Pound	Barite	0.005	0.008	0.009	0.02	0.02	0.03	0.04	0.04	0.05	...
Pound	Bauxite (crude)	0.006	0.007	0.0075	0.023	0.025	0.034	0.041	0.04	0.037	0.039
Kg	Bismuth	1.80	3.19	4	8.41	7.72	3.38	2.50	2.30	1.40	1.75
Pound	Cadmium	1.80	2.33	2.58	4.09	3.36	2.45	2.84	1.93	1.11	0.93
Pound	Cobalt	1.72	1.68	1.65	3.75	4	13.20	25	25	17.26	6.05
Pound	Columbium	0.30	0.66	0.90	1.80	1.90	5.12	6	5	7	...
Pound	Copper (London)	0.24	0.46	0.59	0.93	0.56	0.62	0.99	0.79	0.67	0.72
Pound	Chromium	0.02	0.09	0.16	0.50	0.50	0.40	0.48	0.47	0.38	...
Pound	Fluorite (97% CaF$_2$)	0.02	0.027	0.03	0.04	0.05	0.05	0.08	0.09	0.09	...
Troy ounce	Gold	35	35	35	159.74	161.49	193.55	612.56	459.64	376	424.51
Pound	Ilmenite	0.10	0.24	0.26	0.32	0.39	0.46	0.57	0.69	0.69	...
Kg	Iron ore	0.003	0.01	0.016	0.019	0.023	0.019	0.027	0.024	0.027	0.029
Pound	Lead	0.15	0.14	0.14	0.27	0.19	0.30	0.41	0.33	0.25	0.19
Pound	Lithium (metallic)	12	10.12	9.50	9.38	11.10	13.20	17.15	20.65	20.65	...
Pound	Magnesium	0.20	0.30	0.35	0.75	0.82	1.01	1.25	1.34	1.34	1.43
Kg	Manganese	0.06	0.07	0.08	0.11	0.14	0.14	0.16	0.17	0.16	0.14
Pound	Mercury	1.11	4.62	8	3.53	1.71	1.73	5.24	5.49	4.96	4.14
Pound	Molybdenum (oxide)	0.80	1.42	1.77	2.43	2.95	7.50	9.20	5.15	5.55	5.90
Pound	Nickel	0.35	0.63	0.79	1.74	2.07	2.09	3.42	3.43	3.20	3.20
Kg	Phosphated rocks (PLB grade)	0.006	0.009	0.01	0.05	0.07	0.04	0.05	0.06	0.05	0.04
Onza Troy	Platinum	60	110.37	140	190	164.23	237	439	475	475	417.50
Kg	Potassium (sulphate)	0.03	0.10	0.16	0.30	0.37	0.39	0.62	0.70	0.68	0.62
Pound	Rutile	0.10	0.21	0.28	0.37	0.44	0.51	0.63	0.75	0.75	...
Pound	Selenium	2	4.43	6	19.19	22	18	12.66	4.38	3.53	4.70
Onza Troy	Silver	0.71	1.10	1.29	4.71	4.42	5.40	20.63	10.52	7.95	11.44
Kg	Sulphur	0.02	0.027	0.03	0.06	0.07	0.05	0.09	0.11	0.11	...
Pound	Tantalum	5.80	7.31	8	16	16.50	26.50	27.50	40	25	...
Pound	Tellurium	1.75	4.27	6	8.34	9.28	20	19.77	14	10	...
Pound	Thorium	0.08	0.11	0.12	0.08	0.09	0.10	0.23	0.18	0.19	...
Pound	Tin	0.78	1.42	1.77	3.72	3.12	5.85	7.61	6.50	5.81	5.91
Kg	Tungsten (90% WO$_3$)	0.05	0.035	0.03	0.09	0.09	0.14	0.14	0.16	0.12	0.09
Pound	Uranium (20%)	1.10	2.52	3.50	1.50	3.50	3.25	...	...	...	...
Pound	Vanadium (metallic)	1.10	1.13	1.15	2.72	3.06	3.05	4.04	3.35	3.65	...
Pound	Zinc	0.13	0.137	0.14	0.56	0.34	0.27	0.35	0.38	0.34	0.35

Source:
1. U. S. Department of the Interior, "Minerals Yearbook - vol. I Metals, Minerals and Fuels", various issues.
2. UNCTAD, "Monthly Commodity Price Bulletin", vol. IV, No. 5, May 1984.
3. World Bureau of Metal Statistics, "World Metal Statistics", vol. 37, No. 4, April 1984.
4. Metal Bulletin, April 19, 1984.
5. The World Bank, "Price Prospects for Major Primary Commodities", Report No. 814/82, July 1982.

PROJECTION OF MINERAL CONSUMPTION IN THE YEAR 2000

(Consumption per capita in kgs, total consumption in units of metric tons (MT))

Unit	Minerals	Projection of consumption per capita				Projection of total consumption					
		Latin America Developing countries	Other Developing countries	Developed countries	Centrally-planned economy countries	Latin America Developing countries	Other Developing countries	Developed countries	Centrally-planned economy countries	World total	Annual rate 1980-2000 %
MT	Antimony	31.91	16.07	47.01	31.91	18 189	38 938	42 638	61 586	161 351	4.68
Thousands MT	Asbestos	1.10	0.32	2.55	1.91	629	773	2 315	3 691	7 408	2.09
Thousands MT	Barite	3.78	2.84	4.18	2.18	2 154	5 273	3 791	5 478	16 696	4.03
Thousands MT	Bauxite	82.30	70.23	90.93	61.73	46 911	170 171	82 474	119 139	418 695	7.79
MT	Bismuth	1.27	0.67	1.88	1.40	724	1 620	1 703	2 699	6 746	3.45
MT	Cadmium	7.80	3.63	17.23	11.69	4 446	8 795	15 628	22 566	51 435	5.70
(Gr)	Cobalt	14.00	6.66	33.96	18.10	7 980	16 139	30 800	34 933	89 852	5.18
Thousands MT	Copper	2.00	1.41	7.95	3.00	1 140	1 658	7 211	5 790	15 799	2.62
Thousands MT	Chromium	4.01	3.92	6.43	4.37	2 286	9 501	5 832	8 434	26 053	5.05
Thousands MT	Fluorite	2.07	2.04	2.60	2.33	1 179	4 936	2 358	4 505	12 978	5.23
MT	Gold	0.37	0.16	0.82	0.56	213	396	748	1 085	2 442	3.58
Thousands MT	Ilmenite	1.36	1.33	2.20	1.49	778	3 230	1 995	2 877	8 880	5.73
Thousands MT	Iron ore	172.52	89.86	381.22	258.77	98 336	217 730	345 767	499 426	1 161 259	4.21
Thousands MT	Lead	1.07	0.79	4.24	1.18	610	1 914	3 846	2 277	8 647	2.39
MT	Lithium	3.10	2.60	6.84	4.64	1 767	6 300	6 204	8 955	23 226	6.48
MT	Magnesium	0.12	0.07	0.25	0.23	68	170	227	444	909	5.99
Thousands MT	Manganese	11.09	8.45	11.09	11.09	6 321	20 474	10 059	21 404	58 258	3.98
MT	Mercury	1.73	1.01	3.81	3.44	984	2 438	3 453	6 647	13 522	3.63
Thousands MT	Molybdenum	0.06	0.01	0.13	0.09	34	24	118	174	350	6.06
Thousands MT	Nickel	0.32	0.08	0.70	0.47	182	190	635	907	1 914	5.06
Thousands MT	Phosphated rocks	57.04	56.01	126.04	85.55	32 513	135 705	114 318	165 111	447 647	6.14
MT	Platinum	0.15	0.14	0.15	0.15	88	339	140	299	866	5.77
Thousands MT	Potassium	9.66	4.99	21.35	14.49	5 506	12 091	19 364	27 966	64 927	4.14
MT	Rutile	0.26	0.24	0.99	0.26	146	575	256	500	1 477	6.16
MT	Selenium	1.02	0.30	2.25	1.53	581	727	2 041	2 953	6 302	6.16
MT	Silver	2.27	1.46	5.01	3.41	1 294	3 533	4 548	5 573	15 948	2.15
MT	Tellurium	0.28	0.20	0.61	0.41	160	485	553	791	1 989	7.06
Thousands MT	Tin	0.05	0.02	0.14	0.07	27	50	123	131	331	2.01
MT	Tungsten	14.75	7.10	31.99	20.49	8 407	17 195	29 015	39 546	94 163	2.88
Thousands MT	Uranium	0.03	0.01	0.06	-	17	24	56	39	97	4.03
Thousands MT	Vanadium	0.02	0.02	0.03	0.03	13	56	23	62	154	7.85
Thousands MT	Zinc	1.61	0.99	5.12	2.00	920	2 399	4 644	3 860	11 823	3.31

Source: 1. See table 9 of the Statistical Appendix.
2. Population: United Nations, Monthly Bulletin of Statistics, Vol. XXXVIII, No. 5, May 1984.

Table 21

PROJECTION OF THE BALANCE PRODUCTION - MINERAL RESERVES IN THE YEAR 2000
(Units of metric tons (MT)

Unit	Minerals	Annual growth rates (%) Past 1960-1980	Projected	World Total production Year 2000	Average 1980-2000	Reserves to production. Year of exhaustion
Thousands MT	Antimony	-	4.68	161	113	2 018
Thousands MT	Asbestos	4.08	2.09	7 408	6 155	2 000
Thousands MT	Barite	-	4.03	16 696	11 649	2 000
Thousands MT	Bauxite	6.24	7.79	418 695	256 010	2 071
MT	Bismuth	-	3.45	6 746	5 083	2 000
MT	Cadmium	-	5.70	51 435	34 208	2 000
Thousands MT	Chromium	4.05	5.05	26 053	17 891	2 177
Thousands MT	Cobalt	4.33	5.18	90	61	2 040
Thousands MT	Copper	3.35	2.62	15 799	12 610	2 025
Thousands MT	Fluorite	4.88	5.23	12 978	8 830	2 014
MT	Gold	-	3.58	2 442	1 825	2 000
Thousands MT	Ilmenite	4.98	5.73	8 880	5 897	2 046
Thousands MT	Iron ore	3.78	4.21	1 161 259	835 117	2 122
Thousands MT	Lead	2.10	2.39	8 647	7 017	2 002
MT	Lithium	-	6.48	23	15	2 127
Thousands MT	Magnesium	-	5.99	909	596	-
Thousands MT	Manganese	3.43	3.98	58 258	42 477	2 023
MT	Mercury	-	3.63	13 522	10 072	2 000
Thousands MT	Molybdenum	5.04	6.06	350	229	2 021
Thousands MT	Nickel	3.99	5.06	1 914	1 313	2 032
Thousands MT	Phosphated rocks	6.07	6.14	447 647	291 830	2 223
MT	Platinum	8.73	5.77	866	574	2 044
Thousands MT	Potassium	5.77	4.14	64 927	46 891	2 173
Thousands MT	Rutile	7.80	6.16	1 477	962	-
MT	Selenium	-	6.16	6 302	4 103	2 033
MT	Silver	-	2.15	15 948	13 185	1 998
MT	Tellurium	-	7.06	1 989	1 248	2 028
Thousands MT	Tin	1.07	2.01	331	276	2 015
MT	Tungsten	2.72	2.88	94 163	73 741	2 015
Thousands MT	Uranium	4.39	4.03	97	70	2 017
Thousands MT	Vanadium	8.48	7.85	154	94	2 149
Thousands MT	Zinc	3.16	3.31	11 823	8 994	2 006

		Latin America				
Unit	Minerals	Projected annual rate(%) (Base 1960-1980)	Production		Share in world production in the year 2000	Reserves to production Year of exhaustion
			Year 2000	Average 1980-2000		
Thousands MT	Antimony	2.48	31	22	19.2	2 009
Thousands MT	Asbestos	5.68	421	277	5.7	2 000
Thousands MT	Barite	-0.34	1 097	768	5.7	2 000
Thousands MT	Bauxite	3.37	46 911	32 886	11.2	2 166
MT	Bismuth	1.37	1 679	1 265	24.9	1 998
MT	Cadmium	5.65	4 935	3 282	9.6	2 001
Thousands MT	Chromium	1.96	516	350	2.0	2 000
Thousands MT	Cobalt	16.82	4	2	4.5	2 002
Thousands MT	Copper	1.07	1 772	1 415	11.2	2 113
Thousands MT	Fluorite	4.76	2 332	1 626	18.0	2 012
MT	Gold	-4.57	31	23	1.3	1 996
Thousands MT	Ilmenite	-	80	40	0.9	2 000
Thousands MT	Iron ore	7.15	281 998	176 262	24.3	2 285
Thousands MT	Lead	2.53	610	593	7.1	2 002
MT	Lithium	20.48	11	7	47.8	2 165
Thousands MT	Magnesium	-	68	44	7.5	-
Thousands MT	Manganese	3.15	5 254	3 650	9.0	2 000
MT	Mercury	11.47	640	477	4.7	1 998
Thousands MT	Molybdenum	10.60	108	61	30.9	2 033
Thousands MT	Nickel	5.71	182	125	9.5	2 171
Thousands MT	Phosphated rocks	12.61	32 513	21 156	7.3	2 050
MT	Platinum	8.33	4	2	0.5	1 995
Thousands MT	Potassium	-	5 506	2 753	8.5	2 003
Thousands MT	Rutile	13.83	6	3	0.4	1 999
MT	Selenium	6.70	1 164	758	18.5	2 055
MT	Silver	0.58	3 775	3 121	23.7	1 997
MT	Tellurium	2.64	160	100	8.0	2 012
Thousands MT	Tin	2.35	56	45	16.9	2 015
MT	Tungsten	2.97	7 820	5 550	8.3	2 000
Thousands MT	Uranium	25.29	17	10	17.5	2 012
Thousands MT	Vanadium	18.25	13	7	8.4	2 011
Thousands MT	Zinc	0.48	1 011	795	8.6	2 000

		Other Developing Countries				
		Projected	Production		Share in world	Reserves to
Unit	Minerals	annual rate(%) (Base 1960-1980)	Year 2000	Average 1980-2000	production in the year 2000	production Year of exhaustion
Thousands MT	Antimony	6.13	23	16	14.3	2 000
Thousands MT	Asbestos	2.28	617	418	8.3	2 000
Thousands MT	Barite	1.71	4 725	3 296	24.6	2 000
Thousands MT	Bauxite	11.41	170 171	97 596	40.6	2 089
MT	Bismuth	7.43	390	294	5.8	1 998
MT	Cadmium	11.61	9 000	6 000	17.5	1 999
Thousands MT	Chromium	7.74	9 501	6 881	36.5	2 128
Thousands MT	Cobalt	-0.65	20	17	22.2	2 112
Thousands MT	Copper	0.94	2 306	1 840	14.6	2 060
Thousands MT	Fluorite	10.66	4 272	2 269	32.9	2 000
MT	Gold	7.90	325	243	13.3	1 997
Thousands MT	Ilmenite	13.98	3 230	2 157	36.4	2 011
Thousands MT	Iron ore	-3.56	34 068	24 500	2.9	2 000
Thousands MT	Lead	2.23	490	482	5.6	2 000
MT	Lithium	-	6	4	26.1	2 140
Thousands MT	Magnesium	29.28	170	112	18.7	-
Thousands MT	Manganese	3.15	7 671	4 900	13.2	2 000
MT	Mercury	1.68	1 367	1 018	10.1	1 997
Thousands MT	Molybdenum	21.82	24	13	6.9	2 012
Thousands MT	Nickel	0.96	190	130	9.9	2 243
Thousands MT	Phosphated rocks	5.99	135 705	92 131	30.3	2 502
MT	Platinum	-	-	-	-	-
Thousands MT	Potassium	14.45	12 091	8 578	18.6	2 028
Thousands MT	Rutile	10.39	575	295	38.9	-
MT	Selenium	7.26	1 622	1 056	25.7	2 050
MT	Silver	-4.34	366	303	2.3	1 997
MT	Tellurium	9.33	840	526	42.3	2 043
Thousands MT	Tin	-1.48	109	93	32.9	2 034
MT	Tungsten	2.73	13 881	10 990	14.7	2 002
Thousands MT	Uranium	4.23	21	15	21.7	2 000
Thousands MT	Vanadium	4.89	6	4	3.9	2 004
Thousands MT	Zinc	4.56	798	710	6.7	2 000

(Cont. Table 21)

Unit	Minerals	Projected annual rate(%) (Base 1960-1980)	Developed Countries		Share in world production in the year 2000	Reserves to production Year of exhaustion
			Production			
			Year 2000	Average 1980-2000		
Thousands MT	Antimony	4.00	46	32	28.6	2 006
Thousands MT	Asbestos	3.57	3 973	3 300	53.6	2 000
Thousands MT	Barite	8.68	7 792	5 435	40.6	2 000
Thousands MT	Bauxite	7.11	146 436	85 904	35.0	2 047
MT	Bismuth	4.77	4 158	3 133	61.6	1 998
MT	Cadmium	4.77	33 000	22 000	64.2	2 000
Thousands MT	Chromium	2.35	5 837	3 773	22.4	2 590
Thousands MT	Cobalt	10.17	31	19	34.4	2 000
Thousands MT	Copper	5.11	7 211	5 755	45.6	2 007
Thousands MT	Fluorite	5.08	4 243	3 485	32.7	2 030
MT	Gold	3.54	1 572	1 175	64.4	1 997
Thousands MT	Ilmenite	0.49	2 693	1 788	30.3	2 129
Thousands MT	Iron ore	2.97	345 767	283 453	29.8	2 096
Thousands MT	Lead	5.71	5 649	4 465	65.3	2 003
MT	Lithium	1.12	6	4	26.1	2 043
Thousands MT	Magnesium	-0.24	227	149	25.0	6 463
Thousands MT	Manganese	5.78	23 929	16 591	41.1	2 038
MT	Mercury	4.45	7 160	5 333	53.0	1 998
Thousands MT	Molybdenum	3.29	154	107	44.0	2 025
Thousands MT	Nickel	3.39	635	436	33.2	2 021
Thousands MT	Phosphated rocks	3.46	114 318	75 049	25.5	2 137
MT	Platinum	5.06	298	261	34.4	2 096
Thousands MT	Potassium	1.08	19 364	13 539	29.8	2 251
Thousands MT	Rutile	0.59	396	337	26.8	-
MT	Selenium	3.99	2 041	1 329	32.4	2 029
MT	Silver	3.75	8 148	6 736	51.1	1 997
MT	Tellurium	5.30	553	347	27.8	2 032
Thousands MT	Tin	3.68	35	29	10.6	2 005
MT	Tungsten	3.85	31 629	23 243	33.6	2 006
Thousands MT	Uranium	2.70	59	45	60.8	2 025
Thousands MT	Vanadium	4.55	48	29	31.2	2 263
Thousands MT	Zinc	4.18	7 617	5 739	64.4	2 010

(Conclusion Table 21)

Unit	Minerals	Projected annual rate(%) (Base 1960-1980)	Production Year 2000	Production Average 1980-2000	Share in world production in the year 2000	Reserves to production Year of exhaustion
			Centrally planned economy countries			
Thousands MT	Antimony	6.29	61	43	37.9	2 038
Thousands MT	Asbestos	-0.01	2 397	1 992	32.4	2 001
Thousands MT	Barite	3.45	3 082	2 150	16.1	2 000
Thousands MT	Bauxite	8.04	55 177	39 624	13.2	2 000
MT	Bismuth	1.15	519	391	7.7	1 998
MT	Cadmium	0.59	4 500	3 000	8.7	1 998
Thousands MT	Chromium	5.38	10 199	6 887	39.1	2 010
Thousands MT	Cobalt	11.79	35	21	38.9	2 027
Thousands MT	Copper	4.66	4 510	3 600	28.6	2 000
Thousands MT	Fluorite	1.36	2 131	1 450	16.4	2 000
MT	Gold	3.20	514	384	21.0	2 000
Thousands MT	Ilmenite	13.34	2 877	1 912	32.4	2 010
Thousands MT	Iron ore	5.38	499 426	350 902	43.0	2 070
Thousands MT	Lead	2.97	1 898	1 477	22.0	1 999
MT	Lithium	-	-	-	-	-
Thousands MT	Magnesium	8.81	444	291	48.8	-
Thousands MT	Manganese	2.95	21 404	17 336	36.7	2 020
MT	Mercury	2.67	4 355	3 244	32.2	2 000
Thousands MT	Molybdenum	8.49	64	48	18.2	1 999
Thousands MT	Nickel	7.72	907	622	47.4	2 019
Thousands MT	Phosphated rocks	8.43	165 111	103 494	36.9	2 071
MT	Platinum	8.98	564	311	65.1	2 000
Thousands MT	Potassium	4.57	27 966	22 021	43.1	2 203
Thousands MT	Rutile	19.16	500	327	33.9	-
MT	Selenium	9.19	1 475	960	23.4	2 000
MT	Silver	2.45	3 659	3 025	22.9	1 999
MT	Tellurium	9.20	436	275	21.9	2 000
Thousands MT	Tin	6.67	131	109	39.6	2 003
MT	Tungsten	2.28	40 833	33 958	43.4	2 029
Thousands MT	Uranium	-	-	-	-	-
Thousands MT	Vanadium	9.65	87	54	56.5	2 132
Thousands MT	Zinc	1.90	2 397	1 750	20.3	2 000

Source: See Tables 3 and 9 of the Statistical Appendix.

Table 22

LATIN AMERICA: ESTIMATE OF THE VALUE OF MINING ACTIVITY
IN THE YEAR 2000
(Millions of 1975 dollars)

Minerals	Price per metric ton (1975 dollars)	Production 1980	Production 2000	Consumption 2000	Imports 2000	Exports 2000
Antimony	4 012	78	124	72	-	52
Asbestos	790	111	332	497	165	-
Barite	44	52	48	95	47	-
Bauxite	23	579	1 079	1 079	-	-
Bismuth	18 541	24	31	13	-	18
Cadmium	9 017	9	44	40	-	4
Cobalt	8 267	14	33	66	33	-
Copper	2 050	3 300	3 633	2 337	-	1 296
Chromium	1 102	392	569	2 519	1 950	-
Fluorite	88	81	205	104	-	101
Gold	5 136 334	406	159	1 094	935	-
Ilmenite	705	-	56	548	492	-
Iron ore	19	1 760	5 358	1 868	-	3 490
Lead	595	220	363	363	-	-
Lithium	20 679	81	227	41	-	186
Magnesium	1 653	-	112	112	-	-
Manganese	110	311	578	695	117	-
Mercury	7 782	1	5	8	3	-
Molybdenum	5 357	75	579	182	-	397
Nickel	3 836	257	698	698	-	-
Phosphated rocks	50	151	1 626	1 626	-	-
Platinum	6 109 325	3	24	538	514	-
Potassium	300	8	1 652	1 652	-	-
Rutile	816	1	5	119	114	-
Selenium	42 307	13	7	4	-	3
Silver	151 447	510	572	196	-	376
Tellurium	18 387	2	3	3	-	-
Tin	8 201	295	459	221	-	238
Tungsten	90	1	1	1	-	-
Uranium	3 307	1	2	2	-	-
Vanadium	5 997	3	78	78	-	-
Zinc	1 235	1 135	1 249	1 136	-	113
Totals		9 874	19 911	19 007	4 370	6 274

Source: See Tables 5, 10 and 19 of the Statistical Appendix.

157

Table 23

LATIN AMERICA: DISTRIBUTION OF ESTIMATED MINERAL PRODUCTION IN THE YEAR 2000
(Millions of 1975 dollars)

Minerals	Argentina	Bolivia	Brasil	Chile	Colombia	Cuba	Jamaica	Mexico	Peru	Dominican Republic	Venezuela	Other Countries	Latin America
Antimony		77						37	10				124
Asbestos			273		23			13			23		332
Barite	1		2	4				21	12			8	48
Bauxite			313				448					318	1 079
Bismuth		9						13	9				31
Cadmium			5		4			23	11			1	44
Chromium			500			69							569
Cobalt			11			16						6	33
Copper	55	5	18	2 320				472	727			36	3 633
Fluorite	40		6					159					205
Gold		1	40	4	8			86		5		15	159
Ilmenite			56										56
Iron Ore	54	1 044	3 000	161	54	375		161	134		375		5 358
Lead	29	18	47					138	131				363
Lithium			5	222									227
Magnesium			112										112
Manganese	35	92	386	3				52				10	578
Mercury								4				1	5
Molybdenum	9			510				17	20			23	579
Nickel			49		14	412				174	24	25	698
Phosphated rocks			1 236					73	317				1 626
Platinum					24								24
Potassium			743	901								8	1 652
Rutile			5										5
Silver	8	32	45	37				246	190	3		11	572
Selenium				4				1	2				7
Tellurium				1					2				3
Tin	18	317	111					9	4				459
Tungsten		1											1
Uranium			1					1					2
Vanadium				61							17		78
Zinc	37	75	231					275	631				1 249
Totals	286	1 671	7 195	4 228	127	872	448	1 801	2 200	182	439	462	19 911

TABLE 24

LATIN AMERICA: DISTRIBUTION OF ESTIMATED MINERAL CONSUMPTION IN THE YEAR 2000
(Millions of 1975 dollars)

Minerals	Argentina	Bolivia	Brasil	Chile	Colombia	Cuba	Jamaica	Mexico	Peru	Dominican Republic	Venezuela	Other Countries	Latin America
Antimony	7	1	27	1	3			22	2		2	7	72
Asbestos	25		234	17	30	15	5	94	17		20	40	497
Barite	5		12	20				29	5		4	20	95
Bauxite	108		462				63	205			113	128	1 079
Bismuth	1		1					11					13
Cadmium								27					40
Chromium	88		1 435	151		63		403	75		88	216	2 519
Cobalt	16		35			3		10			2		66
Copper	200	35	958	290				526	105		2	223	2 337
Fluorite	4		20	3				64			3	7	104
Gold	55	20	383	110	58			153	160	16	66	73	1 094
Ilmenite	44		260	14		14		110	27		41	38	548
Iron Ore	152	40	593	140	84	112		374	93		140	140	1 868
Lead	45	15	120					105	33			45	363
Lithium	6		18	5				10				2	41
Magnesium	5		47					60					112
Manganese	35	21	361	14	28	20		125	21		25	45	695
Mercury	1		4		1			2					8
Molybdenum	10		79	18				50	15			10	182
Nickel	65	33	316	50	41	126	16	136		42	28	9	698
Phosphated rocks	86		758		65	35		374	50		50	130	1 626
Platinum					54				387			11	538
Potassium	66	30	760	110	107	100	17	231	50		50	131	1 652
Rutile	11		55	3		3		24			10	7	119
Silver	24	7	73		15			39	20		10	8	196
Selenium	2							2					4
Tellurium	2												3
Tin	16	31	104					42	1			20	221
Tungsten	1		1						8				1
Uranium													2
Vanadium	5		42	8	3			15			5		78
Zinc	91	80	411					284	85		75	110	1 136
Totals	1 176	313	7 583	954	489	491	101	3 527	1 163	58	732	1 420	18 007

Source: See Table 22 of the Statistical Appendix.

Table 25

LATIN AMERICA: NEW RESERVES REQUIRED FOR SELF-SUFFICIENCY IN MINERALS IN THE YEAR 2000

| Minerals | Breakdown of reserves in 1981, in percentages | | | | | | Average production required 1980-2000 | | Reserves required |
	Argentina	Brazil	Colombia	Mexico	Peru	Other countries	Millions of dollars	Thousands of MT	Thousands of MT
Asbestos	-	82	7	4	-	7	304	385	7 700
Barite	-	-	-	59	41	-	73	1 670	33 409
Cobalt	-	67	-	-	-	33	40	5	100
Chromium	-	100	-	-	-	-	1 455	1 320	26 415
Ilmenite	-	100	-	-	-	-	274	388	7 773
Manganese	42	55	-	3	-	-	503	4 373	91 455
Mercury	-	-	-	100	-	-	5	1	20
Gold	-	-	-	100	-	-	750	0.2	3
Platinum	-	-	100	-	-	-	270	0.04	1
Rutile	-	99	-	-	-	1	60	1.42	28

Source: 1. See table 22 of the Statistical Appendix.
2. Federal Institute for Geosciences and Natural Resources, "Regional Distribution of Mining Production and Reserves of Mineral Commodities in the World, Hannover, January 1982.
3. Salas, Guillermo, "Preliminary Study on Mineral Resources of Latin America", Mexico 1979.

ECLAC publications

ECONOMIC COMMISSION FOR LATIN AMERICA AND THE CARIBBEAN
Casilla 179-D   Santiago, Chile

PERIODIC PUBLICATIONS

CEPAL Review

CEPAL Review first appeared in 1976 as part of the Publications Programme of the Economic Commission for Latin America and the Caribbean, its aim being to make a contribution to the study of the economic and social development problems of the region. The views expressed in signed articles, including those by Secretariat staff members, are those of the authors and therefore do not necessarily reflect the point of view of the Organization.

CEPAL Review is published in Spanish and English versions three times a year.

Annual subscription costs for 1988 are US$ 16 for the Spanish version and US$ 18 for the English version. The price of single issues is US$ 6 in both cases.

Estudio Económico de América Latina y el Caribe			Economic Survey of Latin America and the Caribbean	
1980,	664 pp.		1980,	629 pp.
1981,	863 pp.		1981,	837 pp.
1982, vol. I	693 pp.		1982, vol. I	658 pp.
1982, vol. II	199 pp.		1982, vol. II	186 pp.
1983, vol. I	694 pp.		1983, vol. I	686 pp.
1983, vol. II	179 pp.		1983, vol. II	166 pp.
1984, vol. I	702 pp.		1984, vol. I	685 pp.
1984, vol. II	233 pp.		1984, vol. II	216 pp.
1985,	672 pp.		1985,	660 pp.
1986,	734 pp.			
1987,	692 pp.			

(Issues for previous years also available)

Anuario Estadístico de América Latina y el Caribe/
Statistical Yearbook for Latin America and the Caribbean (bilingüe)

1980, 617 pp.
1981, 727 pp.
1983 (covers 1982/1983) 749 pp.

1984, 761 pp.
1985, 792 pp.
1986, 782 pp.
1987, 714 pp.
1988, (en prensa)

(Issues for previous years also available)

ECLAC Books

1 *Manual de proyectos de desarrollo económico,* 1958, 5ª ed. 1980, 264 pp.

1 *Manual on economic development projects,* 1958, 2nd. ed. 1972, 242 pp.

2 *América Latina en el umbral de los años ochenta,* 1979, 2ª ed. 1980, 203 pp.

3 *Agua, desarrollo y medio ambiente en América Latina,* 1980, 443 pp.

4 *Los bancos transnacionales y el financiamiento externo de América Latina. La experiencia de Perú. 1965-1976,* por Robert Devlin, 1980, 265 pp.

4 *Transnational banks and the external finance of Latin America: the experience of Peru,* 1985, 342 pp.

5 *La dimensión ambiental en los estilos de desarrollo de América Latina,* por Osvaldo Sunkel, 1981, 2ª ed. 1984, 136 pp.

6 *La mujer y el desarrollo: guía para la planificación de programas y proyectos,* 1984, 115 pp.

6 *Women and development: guidelines for programme and project planning,* 1982, 3rd. ed. 1984, 123 pp.

7 *Africa y América Latina: perspectivas de la cooperación interregional,* 1983, 286 pp.

8 *Sobrevivencia campesina en ecosistemas de altura,* vols. I y II, 1983, 720 pp.

9 *La mujer en el sector popular urbano. América Latina y el Caribe,* 1984, 349 pp.

10 *Avances en la interpretación ambiental del desarrollo agrícola de América Latina,* 1985, 236 pp.

11 *El decenio de la mujer en el escenario latinoamericano,* 1985, 216 pp.

11 *The decade for women in Latin America and the Caribbean: background and prospects,* 1987, 215 pp.

12 *América Latina: sistema monetario internacional y financiamiento externo,* 1986, 416 pp.

12 *Latin America: international monetary system and external financing,* 1986, 405 pp.

13 *Raúl Prebisch: Un aporte al estudio de su pensamiento,* 1987, 146 pp.

15 *CEPAL, 40 años (1948-1988),* 1988, 85 pp.

17 *Gestión para el desarrollo de cuencas de alta montaña en la zona andina,* 1988, 187 pp.

18 *Políticas macroeconómicas y brecha externa: América Latina en los años ochenta,* 1989, (en prensa)

19 *CEPAL, Bibliografía 1948-1988,* 1989, 648 pp.

20 *Desarrollo agrícola y participación campesina*, 1989, (en prensa)

21 *Planificación y gestión del desarrollo en áreas de expansión de la frontera agropecuaria en América Latina*, 1989, (en prensa)

22 *Transformación ocupacional y crisis en América Latina*, 1989, (en prensa)

MONOGRAPH SERIES

Cuadernos de la C E P A L

1 *América Latina: el nuevo escenario regional y mundial/**Latin America: the new regional and world setting***, (bilingüe), 1975, 2^a ed. 1985, 103 pp.

2 *Las evoluciones regionales de la estrategia internacional del desarrollo*, 1975, 2^a ed. 1984, 73 pp.

2 ***Regional appraisals of the international development strategy***, 1975, 2nd. ed. 1985, 82 pp.

3 *Desarrollo humano, cambio social y crecimiento en América Latina*, 1975, 2^a ed. 1984, 103 pp.

4 *Relaciones comerciales, crisis monetaria e integración económica en América Latina*, 1975, 85 pp.

5 *Síntesis de la segunda evaluación regional de la estrategia internacional del desarrollo*, 1975, 72 pp.

6 *Dinero de valor constante. Concepto, problemas y experiencias*, por Jorge Rose, 1975, 2^a ed. 1984, 43 pp.

7 *La coyuntura internacional y el sector externo*, 1975, 2^a ed. 1983, 106 pp.

8 *La industrialización latinoamericana en los años setenta*, 1975, 2^a ed. 1984, 116 pp.

9 *Dos estudios sobre inflación 1972-1974. La inflación en los países centrales. América Latina y la inflación importada*, 1975, 2^a ed. 1984, 57 pp.

s/n ***Canada and the foreign firm***, D. Pollock, 1976, 43 pp.

10 *Reactivación del mercado común centroamericano*, 1976, 2^a ed. 1984, 149 pp.

11 *Integración y cooperación entre países en desarrollo en el ámbito agrícola*, por Germánico Salgado, 1976, 2^a ed. 1985, 62 pp.

12 *Temas del nuevo orden económico internacional*, 1976, 2^a ed. 1984, 85 pp.

13 *En torno a las ideas de la CEPAL: desarrollo, industrialización y comercio exterior*, 1977, 2^a ed. 1985, 57 pp.

14 *En torno a las ideas de la CEPAL: problemas de la industrialización en América Latina*, 1977, 2^a ed. 1984, 46 pp.

15 *Los recursos hidráulicos de América Latina. Informe regional*, 1977, 2^a ed. 1984, 75 pp.

15 ***The water resources of Latin America. Regional report***, 1977, 2nd. ed. 1985, 79 pp.

16 *Desarrollo y cambio social en América Latina*, 1977, 2^a ed. 1984, 59 pp.

17 *Estrategia internacional de desarrollo y establecimiento de un nuevo orden económico internacional*, 1977, 3^a ed. 1984, 61 pp.

17 ***International development strategy and establishment of a new international economic order***, 1977, 3rd. ed. 1985, 59 pp.

18 *Raíces históricas de las estructuras distributivas de América Latina*, por A. di Filippo, 1977, 2^a ed. 1983, 64 pp.

19 *Dos estudios sobre endeudamiento externo*, por C. Massad y R. Zahler, 1977, 2^a ed. 1986, 66 pp.

s/n ***United States — Latin American trade and financial relations: some policy recommendations***, S. Weintraub, 1977, 44 pp.

20 *Tendencias y proyecciones a largo plazo del desarrollo económico de América Latina*, 1978, 3^a ed. 1985, 134 pp.

21 *25 años en la agricultura de América Latina: rasgos principales 1950-1975*, 1978, 2ª ed. 1983, 124 pp.

22 *Notas sobre la familia como unidad socioeconómica*, por Carlos A. Borsotti, 1978, 2ª ed. 1984, 60 pp.

23 *La organización de la información para la evaluación del desarrollo*, por Juan Sourrouille, 1978, 2ª ed. 1984, 61 pp.

24 *Contabilidad nacional a precios constantes en América Latina*, 1978, 2ª ed. 1983, 60 pp.

s/n *Energy in Latin America: The Historical Record*, J. Mullen, 1978, 66 pp.

25 *Ecuador: desafíos y logros de la política económica en la fase de expansión petrolera*, 1979, 2ª ed. 1984, 153 pp.

26 *Las transformaciones rurales en América Latina: ¿desarrollo social o marginación?*, 1979, 2ª ed. 1984, 160 pp.

27 *La dimensión de la pobreza en América Latina*, por Oscar Altimir, 1979, 2ª ed. 1983, 89 pp.

28 *Organización institucional para el control y manejo de la deuda externa. El caso chileno*, por Rodolfo Hoffman, 1979, 35 pp.

29 *La política monetaria y el ajuste de la balanza de pagos: tres estudios*, 1979, 2ª ed. 1984, 61 pp.

29 *Monetary policy and balance of payments adjustment: three studies*, 1979, 60 pp.

30 *América Latina: las evaluaciones regionales de la estrategia internacional del desarrollo en los años setenta*, 1979, 2ª ed. 1982, 237 pp.

31 *Educación, imágenes y estilos de desarrollo*, por G. Rama, 1979, 2ª ed. 1982, 72 pp.

32 *Movimientos internacionales de capitales*, por R. H. Arriazu, 1979, 2ª ed. 1984, 90 pp.

33 *Informe sobre las inversiones directas extranjeras en América Latina*, por A. E. Calcagno, 1980, 2ª ed. 1982, 114 pp.

34 *Las fluctuaciones de la industria manufacturera argentina, 1950-1978*, por D. Heymann, 1980, 2ª ed. 1984, 234 pp.

35 *Perspectivas de reajuste industrial: la Comunidad Económica Europea y los países en desarrollo*, por B. Evers, G. de Groot y W. Wagenmans, 1980, 2ª ed. 1984, 69 pp.

36 *Un análisis sobre la posibilidad de evaluar la solvencia crediticia de los países en desarrollo*, por A. Saieh, 1980, 2ª ed. 1984, 82 pp.

37 *Hacia los censos latinoamericanos de los años ochenta*, 1981, 146 pp.

s/n *The economic relations of Latin America with Europe*, 1980, 2nd. ed. 1983, 156 pp.

38 *Desarrollo regional argentino: la agricultura*, por J. Martin, 1981, 2ª ed. 1984, 111 pp.

39 *Estratificación y movilidad ocupacional en América Latina*, por C. Filgueira y C. Geneletti, 1981, 2ª ed. 1985, 162 pp.

40 *Programa de acción regional para América Latina en los años ochenta*, 1981, 2ª ed. 1984, 62 pp.

40 *Regional programme of action for Latin America in the 1980s*, 1981, 2nd. ed. 1984, 57 pp

41 *El desarrollo de América Latina y sus repercusiones en la educación. Alfabetismo y escolaridad básica*, 1982, 246 pp.

42 *América Latina y la economía mundial del café*, 1982, 95 pp.

43 *El ciclo ganadero y la economía argentina*, 1983, 160 pp.

44 *Las encuestas de hogares en América Latina*, 1983, 122 pp.

45 *Las cuentas nacionales en América Latina y el Caribe*, 1983, 100 pp.

45 *National accounts in Latin America and the Caribbean*, 1983, 97 pp.

46 *Demanda de equipos para generación, transmisión y transformación eléctrica en América Latina*, 1983, 193 pp.

47 *La economía de América Latina en 1982: evolución general, política cambiaria y renegociación de la deuda externa*, 1984, 104 pp.

48 *Políticas de ajuste y renegociación de la deuda externa en América Latina*, 1984, 102 pp

49 *La economía de América Latina y el Caribe en 1983: evolución general, crisis y procesos de ajuste*, 1985, 95 pp.

49 *The economy of Latin America and the Caribbean in 1983: main trends, the impact of the crisis and the adjustment processes,* 1985, 93 pp.

50 *La CEPAL, encarnación de una esperanza de América Latina,* por Hernán Santa Cruz, 1985, 77 pp.

51 *Hacia nuevas modalidades de cooperación económica entre América Latina y el Japón,* 1986, 233 pp.

51 *Towards new forms of economic co-operation between Latin America and Japan,* 1987, 245 pp.

52 *Los conceptos básicos del transporte marítimo y la situación de la actividad en América Latina,* 1986, 112 pp.

52 *Basic concepts of maritime transport and its present status in Latin America and the Caribbean,* 1987, 114 pp.

53 *Encuestas de ingresos y gastos. Conceptos y métodos en la experiencia latinoamericana.* 1986, 128 pp.

54 *Crisis económica y políticas de ajuste, estabilización y crecimiento,* 1986, 123 pp.

54 *The economic crisis: Policies for adjustment, stabilization and growth,* 1986, 125 pp.

55 *El desarrollo de América Latina y el Caribe: escollos, requisitos y opciones,* 1987, 184 pp.

55 *Latin American and Caribbean development: obstacles, requirements and options,* 1987, 184 pp.

56 *Los bancos transnacionales y el endeudamiento externo en la Argentina,* 1987, 112 pp.

57 *El proceso de desarrollo de la pequeña y mediana empresa y su papel en el sistema industrial: el caso de Italia,* 1988, 112 pp.

58 *La evolución de la economía de América Latina en 1986,* 1987, 100 pp.

58 *The evolution of the Latin American Economy in 1986,* 1988, 106 pp.

59 *Protectionism: regional negotiation and defence strategies,* 1988, 262 pp.

60 *Industrialización en América Latina: de la "caja negra" al "casillero vacío",* 1989, 176 pp.

61 *Hacia un desarrollo sostenido en América Latina y el Caribe: restricciones y requisitos,* 1989, 94 pp.

62 *La evolución de la economía de América Latina 1987,* 1989, (en prensa)

Cuadernos Estadísticos de la C E P A L

1 *América Latina: relación de precios del intercambio,* 1976, 2ª ed., 1984, 66 pp.

2 *Indicadores del desarrollo económico y social en América Latina,* 1976, 2ª ed. 1984, 179 pp.

3 *Series históricas del crecimiento de América Latina,* 1978, 2ª ed. 1984, 206 pp.

4 *Estadísticas sobre la estructura del gasto de consumo de los hogares según finalidad del gasto, por grupos de ingreso,* 1978, 110 pp. (Agotado, reemplazado por Nº 8)

5 *El balance de pagos de América Latina, 1950-1977,* 1979, 2ª ed. 1984, 164 pp.

6 *Distribución regional del producto interno bruto sectorial en los países de América Latina,* 1981, 2ª ed. 1985, 68 pp.

7 *Tablas de insumo-producto en América Latina,* 1983, 383 pp.

8 *Estructura del gasto de consumo de los hogares según finalidad del gasto, por grupos de ingreso,* 1984, 146 pp.

9 *Origen y destino del comercio exterior de los países de la Asociación Latinoamericana de Integración y del Mercado Común Centroamericano,* 1985, 546 pp.

0 *América Latina: balance de pagos 1950-1984,* 1986, 357 pp.

1 *El comercio exterior de bienes de capital en América Latina,* 1986, 288 pp.

2 *América Latina: Indices de comercio exterior, 1970-1984,* 1987, 355 pp.

3 *América Latina: comercio exterior según la clasificación industrial internacional uniforme de todas las actividades económicas,* 1987, Vol. I, 675 pp; Vol. II, 675 pp.

4 *La distribución del ingreso en Colombia. Antecedentes estadísticos características socioeconómicas de los receptores,* 1988, 156 pp.

Estudios e Informes de la **C E P A L**

1 *Nicaragua: el impacto de la mutación política,* 1981, 2ª ed. 1982, 126 pp.
2 *Perú 1968-1977: la política económica en un proceso de cambio global,* 1981, 2ª ed. 1982, 166 pp.
3 *La industrialización de América Latina y la cooperación internacional,* 1981, 170 pp. (Agotado, no será reimpreso.)
4 *Estilos de desarrollo, modernización y medio ambiente en la agricultura latinoamericana,* 1981, 4ª ed. 1984, 130 pp.
5 *El desarrollo de América Latina en los años ochenta,* 1981, 2ª ed. 1982, 153 pp.
5 *Latin American development in the 1980s,* 1981, 2nd. ed. 1982, 134 pp.
6 *Proyecciones del desarrollo latinoamericano en los años ochenta,* 1981, 3ª ed. 1985, 96 pp.
6 *Latin American development projections for the 1980s,* 1982, 2nd. ed. 1983, 89 pp.
7 *Las relaciones económicas externas de América Latina en los años ochenta,* 1981, 2ª ed. 1982, 180 pp.
8 *Integración y cooperación regionales en los años ochenta,* 1982, 2ª ed. 1982, 174 pp.
9 *Estrategias de desarrollo sectorial para los años ochenta: industria y agricultura,* 1981, 2ª ed. 1985, 100 pp.
10 *Dinámica del subempleo en América Latina. PREALC,* 1981, 2ª ed. 1985, 101 pp.
11 *Estilos de desarrollo de la industria manufacturera y medio ambiente en América Latina,* 1982, 2ª ed. 1984, 178 pp.
12 *Relaciones económicas de América Latina con los países miembros del "Consejo de Asistencia Mutua Económica",* 1982, 154 pp.
13 *Campesinado y desarrollo agrícola en Bolivia,* 1982, 175 pp.
14 *El sector externo: indicadores y análisis de sus fluctuaciones. El caso argentino,* 1982, 2ª ed. 1985, 216 pp.
15 *Ingeniería y consultoría en Brasil y el Grupo Andino,* 1982, 320 pp.
16 *Cinco estudios sobre la situación de la mujer en América Latina,* 1982, 2ª ed. 1985, 178 pp.
16 *Five studies on the situation of women in Latin America,* 1983, 2nd. ed. 1984, 188 pp.
17 *Cuentas nacionales y producto material en América Latina,* 1982, 129 pp.
18 *El financiamiento de las exportaciones en América Latina,* 1983, 212 pp.
19 *Medición del empleo y de los ingresos rurales,* 1982, 2ª ed. 1983, 173 pp.
19 *Measurement of employment and income in rural areas,* 1983, 184 pp.
20 *Efectos macroeconómicos de cambios en las barreras al comercio y al movimiento de capitales: un modelo de simulación,* 1982, 68 pp.
21 *La empresa pública en la economía: la experiencia argentina,* 1982, 2ª ed. 1985, 134 pp.
22 *Las empresas transnacionales en la economía de Chile, 1974-1980,* 1983, 178 pp.
23 *La gestión y la informática en las empresas ferroviarias de América Latina y España,* 1983, 195 pp.
24 *Establecimiento de empresas de reparación y mantenimiento de contenedores en América Latina y el Caribe,* 1983, 314 pp.
24 *Establishing container repair and maintenance enterprises in Latin America and the Caribbean,* 1983, 236 pp.
25 *Agua potable y saneamiento ambiental en América Latina, 1981-1990/Drinking water supply and sanitation in Latin America, 1981-1990* (bilingüe), 1983, 140 pp.
26 *Los bancos transnacionales, el estado y el endeudamiento externo en Bolivia,* 1983, 282 pp.
27 *Política económica y procesos de desarrollo. La experiencia argentina entre 1976 y 1981,* 1983, 157 pp.
28 *Estilos de desarrollo, energía y medio ambiente: un estudio de caso exploratorio,* 1983, 129 pp.
29 *Empresas transnacionales en la industria de alimentos. El caso argentino: cereales y carne,* 1983, 93 pp.
30 *Industrialización en Centro América, 1960-1980,* 1983, 168 pp.
31 *Dos estudios sobre empresas transnacionales en Brasil,* 1983, 141 pp.

72 *La evolución del problema de la deuda externa en América Latina y el Caribe,* 1988, 77 pp.
73 *Agricultura, comercio exterior y cooperación internacional,* 1988, 84 pp.
73 Agriculture external trade and international co-operation, 1989, (en prensa)
75 *El medio ambiente como factor de desarrollo,* 1989, (en prensa)

Serie INFOPLAN: Temas Especiales del Desarrollo

1 *Resúmenes de documentos sobre deuda externa,* 1986, 324 pp.
2 *Resúmenes de documentos sobre cooperación entre países en desarrollo,* 1986, 189 pp.
3 *Resúmenes de documentos sobre recursos hídricos,* 1987, 290 pp.
4 *Resúmenes de documentos sobre planificación y medio ambiente,* 1987, 111 pp.
5 *Resúmenes de documentos sobre integración económica en América Latina y el Caribe,* 1987, 273 pp.
6 *Resúmenes de documentos sobre cooperación entre países en desarrollo, II parte,* 1988, 146 pp.

Publications of the Economic Commission for Latin America and the Caribbean (ECLAC) and those of the Latin American and the Caribbean Institute for Economic and Social Planning (ILPES) can be ordered from your local distributor or directly through:

United Nations Publications
Sales Section, — DC-2-866
New York, NY, 10017
USA

United Nations Publications
Sales Section
Palais des Nations
1211 Geneve 10, Switzerland

Distribution Unit
CEPAL — Casilla 179-D
Santiago, Chile